Space and Society

Series editor

Douglas A. Vakoch, SETI Institute, Mountain View, CA, USA
and California Institute of Integral Studies, San Francisco, CA, USA

More information about this series at http://www.springer.com/series/11929

John Traphagan

Extraterrestrial Intelligence and Human Imagination

SETI at the Intersection of Science, Religion, and Culture

Springer

John Traphagan
Department of Religious Studies
University of Texas at Austin
Austin, TX, USA

ISSN 2199-3882 ISSN 2199-3890 (electronic)
ISBN 978-3-319-10550-5 ISBN 978-3-319-10551-2 (eBook)
DOI 10.1007/978-3-319-10551-2
Springer Cham Heidelberg New York Dordrecht London

Library of Congress Control Number: 2014950415

Printed on acid-free paper

Springer is part of Springer Science+Business Media (www.springer.com)

For my wife Tomoko

Acknowledgments

As always, after writing a book there are a number of people who deserve thanks for their willingness to read chapters, discuss ideas, provide editorial assistance, or give ongoing support. The person I must thank first and foremost is my father, who gave me the idea to write this book while we were chatting on the phone one day. We were talking about the course on religion, science, and the search for extraterrestrial intelligence I was teaching at the University of Texas, and he suggested that the basic content of the course would make a great outline for a book. My son, Julian Traphagan, also provided comments on the book, which I much appreciated given his knowledge of and interest in astronomy. I would also like to thank Doug Vakoch, with whom I have had many interesting conversations and who has been quite supportive of my participation in the world of SETI research over the years. My colleagues in the Department of Religious Studies at the University of Texas also deserve my appreciation for the wonderfully collegial environment they create on a daily basis and their patient tolerance for an odd anthropologist with interests in aliens who sits among them. The ongoing help of my research assistant, Katherine Sanchez, has been greatly appreciated. A quick email with a question usually ends up with a flood of articles. Earlier versions of chapters four and five were published in Douglas Vakoch's edited volume *Archaeology, Anthropology, and Interstellar Communication*, which is part of the NASA History Series (2014). I am grateful for NASA publishing my earlier work and allowing me to reuse the material in significantly altered form for this book. Finally, of course, I must thank my wife Tomoko, son Julian, and daughter Sarah, who are always a source of strength and love.

Contents

Chapter 1
Thinking About Religion and Science

One of science's alleged crimes is revealing that our favorite most reassuring stories about our place in the universe and how we came to be are delusional. Instead what science reveals is a universe much older and much vaster than the tidy anthropocentric proscenium of our ancestors.

— Carl Sagan

The search for extraterrestrial intelligence (SETI) represents one of the most significant crossroads at which the assumptions and methods of scientific inquiry come into direct contact with—and in many cases conflict against—those of religion. Indeed, at the core of SETI is the same question that motivates many interested in religion: What is the place of humanity in the universe? When we look up at the sky and wonder about whether or not we are alone, a set of sub-questions are either tacitly or overtly implicated: Are humans unique in the cosmos? Is life abundant in the universe, or is Earth a special place with a special history? Is humanity significant or insignificant when thought about in relation to the cosmos? Both scientists involved with SETI (and in other areas) and people interested in and dedicated to some types of religions are engaged in contemplating these types of questions, even if their respective approaches and answers differ significantly.

In this book I'm interested in exploring this intersection in order to think about three things: (1) the underlying assumptions, many of which are tacitly based upon cultural values common in American society, that have shaped the ways in which SETI researchers conceptualize the nature of their endeavor and represent ideas about the potential influence contact might have on human civilization, (2) what does contemplation of alien others tell us about ourselves? In what ways does SETI express values that we hold about what makes a group of beings intelligent and civilized and what does that say about how we think about ourselves? And, finally,

J. Traphagan, *Extraterrestrial Intelligence and Human Imagination*,
Space and Society, DOI 10.1007/978-3-319-10551-2_1

(3) to what extent are the occasional claims that SETI is itself a kind of faith-based religion a reasonable, or unreasonable, critique of the field?

In terms of the last of these foci—religion and science within the confines of SETI—there are two ways to think about this. One is related to the question of how contact might influence religious belief among adherents to various religious traditions. In many cases, contact with extraterrestrial intelligence (ETI) has been viewed as being at least problematic for many of the world's religions and might in fact represent the death knell for Christianity, in particular, due to its very strong anthropocentric theology (George 2005). The second, and the area in which I am most interested here, raises a more specific question about the nature of SETI itself and the extent to which it may represent something akin to a religious quest because it shares with religion an emphasis on seeking non-human (and often superior) intelligence, the possibility of gaining wisdom from that intelligence, and pursuit of answers to the ultimate question of humanity's place in the universe (Michaud 2007: 202). I'm cognizant of the discomfort astronomers working in SETI research have with the idea of equating SETI with a religion, and this is not my intent, as will become clear in the last chapter. However, there are unavoidable similarities between SETI and some forms of religion and there is an important cultural question related to the desire to encounter another, and potentially more capable, intelligent being present both in many religions (not all) and in SETI that is worth exploring. Why? Because I think, in part, it is no accident that SETI arose in a cultural context heavily shaped by Christianity and its inherent assumptions about the existence of a higher being and a linear model of history that assumes progress is a component of cultural change. I will return to this issue later.

The second area of focus for this book relates to the manner in which assumptions about human "civilization" and culture have influenced the approach scientists working on SETI take when thinking about both the features of an extraterrestrial intelligence and of our own civilization on Earth. It is common for SETI researchers to contemplate civilization in terms of cultural evolution, an idea that has been critiqued quite deeply within the very anthropological circles from which it emerged over 100 years ago. SETI scientists often comment upon human civilization using terms such as "adolescence" or representing humanity as young in comparison to any alien civilization we might encounter. What is usually missed in this formulation is that the notion of adolescence is itself a cultural product and contains tacit assumptions about the nature of both individual and cultural development and change that index a very linear understanding of human social organization as being universally consistent. This concept is then transmitted to ideas about the nature and development of any alien civilization we might encounter. What we do know is that while cultures evolve (meaning that they change) there is no single linear path that they follow, nor does evolution necessarily represent improvement.

The last area of interest I will pursue here—what SETI tells us not about ET but about *Homo sapiens sapiens*—is complex and moves into questions of a moral nature. Peterson (2013: 14) makes an interesting point in her work on environmental ethics when she notes that humans have a long-standing attraction to other species but despite this, many scientists and philosophers see other animal species as morally irrelevant when thinking about human behavior. The fact is that the ways we contemplate non-human aliens, just like the way we contemplate non-human animals, tells us a great deal about our self-concepts as a species and the ways we intertwine moral ideas with other aspects of thought. Scientists engaged in SETI work from a premise that their job is one of discovery—it is science. However, they display deep commitments to often assumed moral propositions about the importance of contact, the nature of supposedly superior civilizations, and the relative inferiority of human civilizations. In other words, the scientific endeavor is shaped by a cultural context that contains moral propositions and assumptions not only about who ET might be, but also what kind of being *Homo sapiens sapiens* is.

1.1 Foundations for Thinking About SETI: Some Ideas and Assumptions

In the remainder of this chapter, I want to focus on thinking about some basic concepts and ideas associated with SETI and to consider how these are related to cultural values. We will work on defining three very widely—and imprecisely—used terms: science, religion, and culture. Before moving into that discussion, however, it will be helpful to offer a few comments about my own assumptions and ideas when it comes to the nature of both religion and science. I view science as a cultural product. By this I mean that the approach to understanding the world associated with scientific inquiry is a product of a set of assumptions, particularly about the relationship between subjective and objective realms of existence, that have shaped Western scholarship and allowed for the development of the type of empirical data collection and systematic methods of analysis that we normally associate with scientific inquiry. It is important to recognize that both science, and scientists, are embedded in cultural and social contexts that shape the ways in which they ask questions, determine which questions are important, and respond to the more philosophical components of their inquiries. These contexts can also influence the ways in which they interpret empirical data.

Science is a human activity that is closely tied to affluence; it is a luxury item, particularly when it comes to pursuit of questions such as the existence of extraterrestrial intelligence. By luxury, I do not mean that it is extravagant and an example of excess. Rather, science is something that arises when there is sufficient wealth for some people in a society to be occupied in activities well beyond maintenance of their basic survival. This is not to say that scientists do not contribute something profoundly important to human society; rather my point is that science can only exist as an institution in a context that can afford to have certain people working in very specific types of endeavors while others do jobs that support those people. This is true for both physical and social scientists. The capacity to do what I do—get paid to think about the nature of culture, society, religion, SETI—is a product of an affluent society that can afford to have people engaged in thinking about how human social organization functions. The fact that we can afford physical and social sciences is a good thing because it provides a basis for both building new ways of seeing the world around us and also for reflexively contemplating who we are as a group and as a species. But it is still a luxury.

Why is this important? Because although this fact rarely is contemplated by scientists, it does shape the way we think about the significance of our work. This is particularly true when it comes to the significance of SETI. Although I would agree with those involved with SETI research that contact with ETI would represent a major moment in human history, one also should recognize that for a very large part of humanity, the existence of ETI is largely irrelevant—most people's time and energies on Earth are not occupied with contemplating alien civilizations but with managing survival in an environment where resources are scarce and very unequally distributed. According to the World Bank, roughly 1.25 billion inhabitants of Earth live in crushing poverty, surviving on less than $1.25 a day. About 2.5 billion people live on less than $2 a day, and approximately 80 % of the planet's population lives on less than $10 per day. Although questions have been raised about the strength of the empirical basis for these data (Chandy and Kharas 2012), the fact remains clear that most of Earth's population lives in conditions ranging from moderate to extreme poverty.

This might not seem like an important issue for a book about research into the search for extraterrestrial intelligence, but it is necessary to recognize the relationship between economic and social factors in allowing people to pursue questions related to this topic or any topic of a scientific, sociological, or philosophical nature. The capacity to explore the cosmos with radio telescopes and to devote a lifetime in pursuit of contact with ETI arises in a socioeconomic and cultural milieu that both generates a distribution of resources necessary for the endeavor to operate and contains cultural values that encourage the perception that this is both interesting and important as an activity—*there is nothing inherently important or interesting about contacting an extraterrestrial intelligence*; *it is interesting because we live in a culture that has come to value the idea of contact with alien intelligence.* However, for the majority of humans the quest for contact with ETI has little relevance to the necessities of procuring the basic goods needed to simply get through each day. In other words, the conditions that allow for the science of SETI are not shared by the majority of humans on Earth and are, in fact, a specific consequence of the development of industrial and postindustrial society that allows for the economics of scientific discovery and generates a cultural context in which the questions associated with SETI become valued and important to many members of the societies in which SETI research is pursued.[1]

My position on religion is that it represents one of the most important areas of human social and imaginative activity and profoundly influences, obviously, the behaviors of many of Earth's inhabitants. Therefore, it is essential for researchers to pursue a deep understanding of the nature, influence, power, and scope of religious

[1]Again, I want to emphasize that I'm not arguing that SETI research is unimportant. I think it's very important, but it must be understood within the context in which it arose and in terms of how it is connected to that context and the values associated with that context.

behaviors and ideas. That said, I'm neither religious nor inclined to believe in the existence of any sort of god. For the most part, I agree with Richard Dawkins (2006: 57) that a survey of human religious traditions allows one to "marvel at the richness of human gullibility." And while I find that gullibility quite fascinating, I also find it rather disturbing that so much of human history has been shaped by belief in, or response to belief in, the presence and power of fictitious characters such as the Abrahamic god that sits at the center of Christian, Islamic, and Jewish theology. In short, my interest in religion does not come from the perspective of being a believer, but rather of a disbeliever who is fascinated about how belief in such ideas shapes human behavior. I tend to be less troubled by some non-Western traditions, such as Buddhism, that are not so inclined to build theological ideas and religious institutions around the worship of a fictional godhead—Buddhism, in fact, lacks a deity. But they remain equally fascinating in their power to generate and promulgate a cosmology that profoundly influences individual and collective behaviors and ideas and motivates action. As a social scientist, I try to set aside these biases concerning religion, but the fact remains that they are there and should be understood by the reader as contributing, even if only subconsciously, to how I understand and define religion.

Having read the previous paragraph, the religiously minded reader may be inclined to put the book down right now, and I understand how that feeling could arise. But I would encourage you to forge on, at least for a bit—I most likely won't disparage religion again for a while, as this book is not a critique of religious ideology and theology. I'll work on that one another time. But it is important for readers to understand my biases and intellectual leanings as I write the pages that follow. Central in my perspective is the conviction that religions, particularly of the Abrahamic ilk, work from cosmologies that make very little sense and that the theological writings supporting and expanding upon those cosmologies are neither scholarship nor research. Indeed, I view theology as little more than writing stories based upon other stories such as those found in the Bible. Theological writing is not based upon evidence, has no grounding in empirical reality, and is antithetical to scholarship, research, and scientific reason. And in the end, a great deal of theology (probably most of it, but I haven't read it all, so I can't make that claim in good conscience) is based upon failed understandings about and faulty observations of the nature of physical and biological reality. Okay, I'll stop.

1.2 Science and Religion

A question I often ask students in my courses related to religion and science is whether or not both are engaged in seeking truth. Students are often somewhat nonplussed at this—I think they assume it's a trick question—but usually come down on the side that yes they share the aim of arriving at truth about the universe, about its origin, future, and make-up. In fact, the question is actually quite difficult to answer. Both science and religion often seek answers to very big questions about the nature of the cosmos. However, the approach taken by each is obviously quite different and the types of answers at which they arrive are distinct. When it comes to big questions such as the origin of the universe, science is generally concerned with what happened and how; religion is concerned with why and what it all means.

Both terms require definition in order to explore any commonalities or differences between them, but I prefer to avoid falling into the trap so nicely elucidated by physicist Richard Feynman (1998) that too precise a definition leads to the kind of intellectual paralysis that philosophers routinely encounter as they try to out-define each other about every word they are using. There are, in fact, quite a few ways to think about what both science and religion are, many of which have both useful and problematic components.

When it comes to science, Feynman noted in his book *The Meaning of it All*, that the term is used imprecisely; the word "science" can refer to a way of seeing the world, a body of knowledge about the world, or the products of that knowledge expressed in the form of technology. A fairly representative definition of how scientists think about their own work can be found in Isaac Asimov's comments in an interview with Bill Moyers during a broadcast in 1988:

> Science does not purvey absolute truth; science is a mechanism. It's a way of trying to improve your knowledge of nature. It's a system for testing your thoughts against the universe and seeing whether they match.

Expressing a similar sentiment, Stuart Firestein (2012: 22) in his book *Ignorance: How it Drives Science*, writes that, "[r]eal science is a revision in progress, always. It proceeds in fits and starts of ignorance." I think most scientists, whether working in natural or social science disciplines, if pressed to contemplate what they do on a daily basis would agree with this notion that science is inherently contingent. Perhaps what most thoroughly identifies the scientific approach is an acceptance of the idea that our understanding of the universe is always susceptible to revision and that whatever conclusions we draw tend to highlight our broader ignorance more than they provide answers to anything. More important than obtaining answers to our questions about the world, the scientist is one who focuses on coming up with the right questions to ask about our world. In this sense, science is a human endeavor that emphasizes the value of seeking understanding through the process of asking well thought-out questions, but it is inherently suspicious of the answers we get to any questions we might ask. This is applicable to both the natural and social sciences.

What we can say about science is that scientists of any stripe generally agree on three main points: (1) good science begins with good questions, and (2) all answers to questions we ask are inherently contingent; therefore (3) our descriptions of the world developed through scientific inquiry are inherently contingent. When an experimental scientist arrives at a result, we can verify that result by running the experiment again to see if that result can be replicated. This does not mean that the scientist has arrived at a permanent and final understanding of that aspect of the world. Rather, it is true in the sense that, based upon our current understanding, the result appears to accurately represent a particular aspect of nature; should a better way of representing that aspect of nature arise, then either (A) the initial result will be invalidated or (B) the scope of that result will be limited. This particular approach does not work very well with observational sciences, such as anthropology or field biology, in which the conditions are constantly changing, thus making replication impossible. But the basic assumption remains that if another scientist studies the same context at some point in the future, the initial observations will likely be revised based upon changing conditions. In other words, the "answers" arrived at through observation are inherently contingent, just like the "answers" arrived at through experimentation.

Scientists may work under the general assumption that a particular theoretical framework within which they are operating is accurate, but they remain generally open, under certain conditions related to the overall paradigmatic structure of what Thomas Kuhn (1962a) describes as normal science, to revision of a particular theory. In many cases, this openness results in a narrowing of the scope of applicability of a theory or in the rethinking of the particular way in which natural or social processes operate given the emergence of new empirical evidence. For example, Darwin's understanding of evolution was based upon the idea that very slow, gradual processes of change lead to the transformation of entire populations and, consequently, the emergence of entirely new species. Over the course of billions of years, this process has led to the kind of biodiversity that we see in nature today and is known as phyletic gradualism, which is seen from the traditional Darwinian perspective as being relatively smooth and occurring at a fairly consistent rate over long periods of time, although that rate can be affected by sudden events that interrupt the flow, such as the catastrophic impact that apparently brought the dinosaurs to their demise.

Unfortunately, the fossil record does not clearly support the kind of incremental change in organisms that phyletic gradualism predicts; in fact, we tend to find various organisms that appear to be related, but for which we often can't find much in the way of interim organisms. There are a couple of ways to respond to this problem. One is to assume that the fossil record is incomplete. Although we can see the connections between different organisms, such as hominids like *Homo habilis*,[2] *Homo ergaster*, and *Homo erectus*, and can identify a relatively linear progression that shows these hominids as descendants of early *australopithecines* and ancestors of modern humans, nature simply does not maintain the fossil record well enough for scientists to identify all of the intervening steps that led from the development and transition from one hominid species to another. In other words, there are gaps in the fossil record that make it difficult for us to track the precise process of gradual morphological change in species that occurred over very long periods of time, but the problem is not with the theory of phyletic gradualism, it is with that lack of complete data to fully support the theory which, nonetheless, seems sound based upon the data we have.

---

[2] There has been debate among scholars about whether *homo habilis* belongs in the genus *Homo* rather than *Australopithecus* due to its brain size, but for our purposes the point remains consistent here.

An alternate response, developed by Niles Eldredge and Stephen Jay Gould (1972, 1977), rests on the idea that the problem is not with the fossil record, but with the theory of phyletic gradualism. Rather than working from the position that the fossil record is incomplete, Eldredge and Gould chose to treat the "gaps" as real data rather than as missing data. As a result, the "gaps" in the fossil record are viewed as accurate representations of the tempo of evolutionary change, which rather than happening smoothly occurs in fits and starts. The basic point of the theory developed by Eldredge and Gould, known as punctuated equilibrium, is that very long periods of relative stasis in the morphology of species are punctuated by brief periods in which rapid changes develop and significant speciation occurs. As Eldredge and Gould write in their ground breaking paper (1972: 84), the "history of evolution is not one of stately unfolding, but a story of homeostatic equilibria, disturbed only 'rarely' (i.e., rather often in the fullness of time) by rapid and episodic events of speciation."

The nature of this debate is usually misunderstood by religious types who are either suspicious of or want to challenge the accuracy of evolution as a way of describing the history of life on Earth. These individuals often make the mistake of arguing that one of the "flaws" of evolutionary theory is the "contradiction" between gradualism and punctuated equilibrium. This idea, like many among fundamentalist Christians (and others), betrays a lack of understanding of science and of theory, rather than a "flaw" within evolutionary theory. Scientists who work in the area of evolution (and most other scientists as well) have no dispute about the basic Darwinian insight that biological change occurs through the process of natural selection—both gradualists and those in favor of punctuated equilibrium agree on this. The disagreement is about how the process of natural selection operates over time.

In fact, the evidence for natural selection is overwhelming and can be seen in many observed processes in nature, such as changes in the distribution of black and white peppered moths during and following the industrial revolution in Manchester, England, in which moths predominately gray with black speckles that were the dominant form of the species were replaced by moths that were largely black. This appears to have been related to pollution in the form of sulfur dioxide emissions from local coal plants that killed lichen on trees or landed on trees with gray bark (Jermain et al. 2009: 29–30). As the environment changed due to the pollution, the gray moths increasingly stood out against the darker background of the tree bark on which they lit, making it much easier for birds to see, and eat, them. By contrast, the black moths became camouflaged against the darker background of the blackened trees, making it more difficult for birds to see them. As the birds ate the moths that they could now see and missed the black moths that blended into the sooty bark, the genes for the gray moths were reduced in the population and those of the black moths expanded, because the black moths had opportunities to reproduce denied to the gray moths as a result of being eaten by birds. Following England's clean air legislation and subsequent reduction in air pollution, the distribution of gray peppered moths in the population increased. This is exactly the process that Darwin describes in his discussion of natural selection and represents solid empirical evidence that what Darwin observed and described about how nature works is accurate.

Nobody from either side of the debate about gradualism and punctuated equilibrium would argue against the idea that the peppered moth example shows anything other than the fact that Darwin was right about the basic process of evolution as occurring through natural selection. What these two camps within evolutionary biology disagree on is how to read the fossil record and, as a result, how to interpret the tempo and flow of evolutionary change. To argue that this represents a fundamental problem with evolutionary theory is equivalent to arguing that because Newton and Einstein have different ideas about how gravitational forces work, the entire notion that gravity exists is flawed. This type of position not only betrays a lack of understanding of both science and the natural world, it is logically untenable because it represents an example of the fallacy known as the inverse error. Those who take this position in essence argue that if gradualism (or punctuated equilibrium) is correct (P), then evolutionary theory is correct (Q); because gradualism (or punctuated equilibrium) may not be correct (not P), evolutionary theory is not correct (therefore not Q). Arguments in this form are logically invalid because they fail to give an acceptable reason to establish the conclusion, even if the initial premise is correct.

Having explored what we mean by science, I now want to turn to a discussion of what we mean when we use the word religion. Like the term science, religion is usually imprecisely defined and can refer to a variety of different realms of human experience. One way to discuss religion is to think of it as a particular way of seeing the world that emphasizes ideas and beliefs associated with concepts such as spirituality, morality, and human/natural origins. The term may also be used to refer to *institutions* that develop, support, and disseminate a particular set of ideas about how to see the world. In general, when adherents to Western-style religions talk about religion, they work from the assumption that the core of being religious is having faith in the basic dogmas and ideas held sacred by others who share the same worldview. This typically means believing in the same god and accepting, on faith, the idea that the tenets of the religion are inherently true.

This way of thinking about religion is by no means universal. In many Asian societies, the emphasis of religious life is far less centered on belief than it is on regular performance of rituals (Traphagan 2004). Japanese, for example, will routinely state that they do not see themselves as religious in the sense that they believe in a deity or have "faith," while also regularly participating in ritual activities associated with both Shinto and Buddhism. Indeed, faith is not considered a prerequisite for participation in religious activities and, unlike Abrahamic religions, belonging is not seen as exclusive. It is quite normal for Japanese to participate in both Shinto and Buddhist rituals and many even choose to get married in a Christian ceremony, all the while expressing absolutely no commitment to any of these religious traditions on the basis of faith in the worldviews, deities, or dogmas associated with them (although Shinto and Buddhism in Japan tend not to have much in the way of dogma in any case).

When scholars have contemplated the nature of religion, the varieties of religious definitions have been quite extensive. The first individual to offer a formal definition of religion was Edward B. Tylor (1920: 424), who argued that the initial step in understanding animism—or the religious behaviors associated with polytheism and characteristic of the "lower races" as he put it—was to define the term at least minimally as the belief in spiritual beings. For Tylor, who influenced the anthropologist James G. Frazer of *Golden Bough* fame, religion arose in prehistoric times as a result of dreams in which people found themselves among both the living and the dead. Tylor tried to imagine what interpretive spin earlier people might have put on this experience, and he concluded that they would reason that some part of the human self existed as an insubstantial entity that complemented the material body and survived death. These ideas evolved into more elaborate notions about spirits that were seen as existing independently of humans. To win favors, humans made offerings to these beings and this is how ritual evolved. As more powerful spirits were imagined, polytheism came into being and eventually monotheism evolved out of that, representing a "higher" and more complex form of religious behavior and organization.

Tylor's definition of religion is important not so much for its imaginative content, but for the fact that he is the first social scientist to attempt to look at the internal logic of religious behavior and belief. He recognizes that religions represent rational systems that, in his view, "primitive" peoples used to make sense of experiences beyond their capacity to explain through other means. In short, religion is a consequence of human reason—like science, it represents a means by which people try to make sense of the world. Tylor saw religious practices as early attempts to explain the world, and specifically to explain life and death, and he further postulated that in the modern world certain remnants of these ideas continued in the form of things like superstitions, which were primitive *survivals* that continued to operate in more advanced societies. These survivals bear similarity with biological vestigial structures, such as the appendix in humans, which have lost most or all of their original function through evolutionary processes.

A second aspect of Tylor's work, one that was common among many anthropologists and other social scientists working in the later nineteenth and early twentieth centuries, is that he was clearly influence by Darwin. When Tylor wrote about "lower races" and "primitives" he was not only displaying the racism common at the time, he was structuring his understanding of religion (and culture more generally) upon assumptions about the idea of cultural evolution expressed in social Darwinism. Many scholars developed ideas of cultural or social evolution in which there were clearly identified stages of cultural development that did not occur at the same rate in all societies, but that were conceived as happening faster for Europeans. Lewis Henry Morgan (1877), who laid the groundwork for the development of anthropology in the US with his study of Iroquoian kinship, identified three stages of cultural evolution: (1) savagery, characterized by use of fire, the bow, and pottery, (2) barbarism, characterized by domestication of animals, agriculture, and metalwork, and (3) civilization, characterized by use of the alphabet and writing. What's important here is that Morgan links social development with technological development and argues that the measure of the advanced state of a society should be based upon its level of technological development, an idea that he expands to include stages of cultural or social development, as well.

I will write more about this later in the book, because it is relevant to the manner in which SETI researchers often think about the possible nature of ETI. For now, what matters is that ideas associating social evolution with technological progress, as well as the attempt to rank societies on the basis of their stage of technological and social development, were abandoned by anthropologists and other social scientists in the twentieth century. And the notion that one religion, or one culture, is in some way more advanced on an evolutionary scale has also been abandoned. But numerous other attempts to define religion in terms of various basic elements have been proffered by scholars. These include ideas that emphasize identifying the distinct qualities of sacred as opposed to the profane realms of human experience and avoiding the temptation to reduce religious experience in ways that allow for explanation through social science or biological frameworks (Eliade 1959), the notion that religion deals with that which ultimately concerns humans (Tillich 1951), Freud's (1950) idea that religion resulted from guilt of sons over killing and eating their father because he had all the luck with women (few people in human history have had a more fertile—and often ridiculous—imagination than Freud), or that religions are systems of symbolic representation that serve the purpose of creating and identifying what is real—a cosmology—and powerful among a given group of people (Geertz 1973).

The list is really quite long and I'm (somewhat unfairly) truncating the arguments of these scholars, as well as leaving out many important scholars who have attempted to define religion; the short of it is that scholars have never really arrived at a definition of religion upon which they can agree. One scholar whose work has been profoundly influential in thinking about religion is Émile Durkheim (1858–1917), who is often accredited as the founder of sociology and who also influenced early theoretical developments in anthropology. Emphasizing the function of religion in human society, Durkheim offers what I think is in many ways the best definition to come along, despite the fact that it is now about 100 years old. Religion, for Durkheim (2008: 47) is: "A unified system of beliefs and practices relative to sacred things, that is to say things set apart and forbidden—beliefs and practices which unite into one single moral community, called a Church, all those who adhere to them." There are several important points here. First, religion is communal; it is something that unites people and holds them together. Second, religion involves specific beliefs (doctrine and dogma in the West) and practices (rituals) that function as glues that hold groups of people together to form that community. Third, religion is not so much about gods and supernatural beings as it is about identifying what matters and what doesn't matter and using those definitions as a means of controlling people and constructing a sense of group identity. In other words, religion is about power and the god of a religion is not in truth a supernatural being, but society itself held together as a moral community that, in many cases, is perceived as being superior to other groups. Durkheim shifts the question away from what religion *is* to what it *does* and in doing this he points out that religion is a social institution that contributes to the structure and function of society and also forms one way in which groups of people define themselves as groups—in terms of morality—in opposition to other groups of people. It also removes the issue of the existence of a god from the realm of the scientific study of religion and allows us to think about what people *do* with their gods and how they make those gods seem real and powerful.

Having discussed both science and religion the question now arises: How do we compare and distinguish these two? At the beginning of this chapter, I noted that science and religion share certain features. They both can be focused on ultimate questions and they both attempt to arrive at explanations, derived from an internal logic, as to how the universe operates and is organized. Religion, of course, moves into other realms such as that of morality and the expression of value judgments, representing a kind of institution quite different from science in that it is overtly focused upon organizing and unifying people into like-minded associations whose behavior is controlled and limited by the dogmas, rules, and ideologies of the institution (although as Kuhn has argued, these things can operate in science as well). Scientists also express value judgments within the context of their work as scientists—claims that contact with ETI will have a profound influence on humanity and change our understanding of ourselves and our place in the universe are value judgments. They are not based upon empirical evidence, because there are no empirical data upon which to develop an analysis and interpretation at this point—we haven't made contact. We will see what happens if that contact actually occurs.

Furthermore, it is important to recognize that scientists live and work within the context of institutional and disciplinary ideological matrices that influence how they think about problems and approach their work. Earlier in this chapter, I noted that Kuhn's concept of normal science allows for a certain openness to alternate ways of thinking which generates opportunities for the development of new theories and new ways of describing the world. But normal science also structures and restricts the ways in which scientists think and the types of questions they ask. In normal science, scientific inquiry—the daily work of scientists—is largely aimed at the articulation of observed phenomena and theoretical frameworks that a given paradigm supplies, rather than the creation of new theories. In other words, scientific inquiry is conducted within the context of a paradigm that shapes and in many cases limits the range of questions that are normally asked. A given paradigm provides a roadmap for thinking that is necessary if scientists are going to advance knowledge, but it also tends to influence and in many cases limit the types of questions that are considered normal and acceptable, thus inhibiting the generation of new and novel theories. The primary mechanism by which this limiting action occurs is peer review, which can place a significant damper on the publication of novel and creative ideas that challenge conventional practice.

Scientists are human beings and, thus, may concern themselves with not only the pursuit of new knowledge, but also the pursuit of prestige and power. Kuhn makes the important observation that as a result of the emphasis within scientific training on linking historical individuals with discovery, the act of discovery itself can become an important personal goal. Kuhn (1962b: 760) argues that "[t]o make a discovery is to achieve one of the closest approximations to a property right that the scientific career affords. Professional prestige is often closely associated with these acquisitions." This can lead, of course, to the types of acrimonious disputes that sometimes arise among scientists over the ownership or independence of a particular discovery or the reasonableness of a competing theory. That said, expressing value judgments and seeking personal gain is neither the function nor aim of science, rather it is a byproduct of the fact that people with similar interests and ideas will both congregate and also attempt to wield power over each other and manage or manipulate the behaviors of peers and competitors. Unlike many religions, however, science is not an institution aimed at social control, even while in some cases normal human proclivities toward seeking personal gain and defending personal prestige may shape the flow and direction in which scientific inquiry unfolds.

Where the most fundamental difference between science and religion lies is in the issue of contingency. While both science and religion are examples of explanatory systems used to make sense of the world, and religion may, as Tylor points out, have begun with some sort of empirical observation as a basis for developing ideas about the nature of that reality or may rely upon observations of events deemed "miracles" as a way of validating a set of beliefs, in modern societies religion tends to work from broad truths that are assumed to be correct and then draws specific conclusions about the world based upon those assumed truths and beliefs. In the case of many religions, these assumptions, from an ideological perspective, are neither open to challenge nor revision, despite the fact that from an empirical perspective they do undergo revision and hybridization due to changing interpretations of basic texts and rituals within the context of shifting cultural frameworks over historical time, as well as the fact that they merge and intermingle with indigenous religious and cultural ideas as religious people move and come into contact with other religious people and ideas in new places (Kapchan and Strong 1999). Clearly, many religions—and particularly those of the Abrahamic tradition—do not work from the assumption that the worldview they present is contingent. Notable exceptions to this are East Asian religions, such as Taoism and Buddhism, that differ significantly from Abrahamic religions in part precisely because they operate from the basic assumption that the universe itself *is* inherently contingent and that the experience of permanence is an illusion.

For religions like Christianity and Islam, the tenets of their basic worldview are conceptualized as being fairly certain or even absolute, while the ideas and interpretations developed from that basic worldview may be understood as changing and contingent to some extent. Thus, both within and among various Christian sects there is an awareness that passages of the Bible can be interpreted in different ways and in Islam there is a long history of interpretation of the Quran in the form of the Sharia, but there is rarely a challenge to the idea of the fundamental verity of the Bible or Quran or the basic assumption that the Trinity is real or that Allah is the only god. Most Christians and Muslims agree (among themselves, at least) that there is a basic, unquestionable truth to their way of seeing the world, even if they don't always agree on how to understand and interpret that truth. In other words, while they may disagree on the specifics of how to interpret the Bible or the Quran, Christians and Muslims generally agree within their own sects—or more precisely, they have faith—that their (idiosyncratic, I must add) perspectives represent some fundamental truth about the universe that is knowable with relative certainty or with complete certainty, depending upon whom you talk to. The knowing that comes with religions like Christianity, Islam, and Judaism—but not with all religions by any means—is grounded not in a well-defined method of generating knowledge that characterizes the empirical sciences, but in faith, which represents a non-rational and unverifiable source of knowledge and truth about the world.[3]

Science approaches understanding and knowledge in an entirely different way. First, science usually begins with specific observations of the world and then attempts to develop theories of underlying principles and processes that explain those observations. In one sense, this has similarities with religion because clearly the same basic response to reality—observation generating explanatory schemas—can be shared between religion and science. Note that this does not mean that a particular schema is correct; it simply means that the starting point for religion and science is sometimes congruent, even if various religious traditions often develop from an initial observation and explanatory schema that is ultimately false. However, unlike religion, science involves the *systematic* study of the world through carefully planned observation in order to generate and organize knowledge that can be tested and can, in some cases, lead to predictions about the universe. Furthermore, scientists work from the basic conviction that it is necessary to *verify* observations before drawing any conclusions about accuracy.

So what does science, more precisely, involve? The basic norms of science are as follows. Science:

1. *Involves gathering of data*—This is understood in a very broad sense that ranges from the type of quantitative data associated with measurement in the natural sciences and some social sciences such as sociology to the types of qualitative data associated with cultural anthropology. Data are empirical in that they are based upon observations of the world that are as unbiased as possible.
2. *Must be objective*—The meaning of "objectivity" is open to debate, and scientists have long understood the notion that we can obtain truly objective data and perform truly objective analysis to be a delusion (Bernard 2011: 4). When we think about an observation as being objective, this does not mean that it should be seen as corresponding to an objective reality that is distinct from human mental activity. Instead, empirical data are collected and interpreted within space and time (Pring 2009: 323), which means that both methods of collection and approaches to interpretation are shaped by cultural context. Observations (and empirical data) represent what might best be understood as a complementary picture of the thing being studied, a picture that operates as a means by which the scientist interprets phenomena. In other words, empirical data are fundamentally symbolic in that they are representations of experience that elicit particular kinds of interpretive responses (Whitehead 1927: 8). However, scientists generally hold that striving for objectivity is a worthwhile endeavor because it forces us to be explicit about

[3] By this, I mean that it provides this source of knowledge and truth from the perspective of believers. I am in no way arguing that the methods associated with Abrahamic religions produce anything more than stories about the world.

our methods and measurements, thus allowing others to identify our errors and improve our understanding of the world (Bernard 2011: 4).

3. *Must be verifiable*—That is, the observations made must have the capacity to be observed by others and confirmed as accurate, although there is a general understanding that in field sciences like anthropology and primatology it may be impossible to actually replicate a particular observation because the subjects of the study, the researcher, and the context are constantly changing.

Did you read the word "truth" in that definition? In fact, as I wrote the above list, as well as the discussion that preceded it, I made it a point to avoid the word "truth." My reason for this is that truth is a very complex concept that, although we often treat it as though it represents universal and unwavering propositions or knowledge, is extremely difficult to pin down in any definitive way without appeal to some type of non-rational concept such as faith. When it comes to science, the fact is that what we are looking at isn't a process of finding truth. As Firestein notes (2012: 28) science doesn't operate along the lines of the proverbial onion in which one strips away layer after layer to get at the truth lurking deep inside. Rather, it is like the expanding ripples that emerge on the surface of a pond after one throws in a rock; the wider they become, the more of what is beyond—the unknown—they manage to touch. However, the most powerful thing that expansion does is to uncover more indications of the extent to which we don't know things. Science rides upon the outer ripple, ever perched on the edge of uncertainty and ignorance, rather than at the inner sanctum of deep understanding about the cosmos (this is in complete contrast to religions like Christianity and Islam, in which each sect within the broader framework of the religion usually perceives of itself as sitting right there at the middle where the rock landed). There is no question that scientific inquiry generates new forms of knowledge and new understandings of the world, but each time we learn something new through science there is an associated portal to ignorance that opens and reminds us that there is much more to be known. In short, science does not provide us with *the answer*; it provides us with contingent answers that primarily function to raise further questions.

As you move with me throughout the remainder of this book, please keep these perspectives of religion and science in mind. Fundamentally, I see both religion and science as products of culture and of a human desire—perhaps need—to situate ourselves in our surroundings and understand our place in the universe. Both science and religion attempt to provide cosmologies, even if the logic of those cosmologies and the manner in which they are constructed is different and, in many cases, inherently at odds. In the next chapter, we will turn to an exploration of how the capacity for humans, or at least Europeans, to imagine other worlds with intelligent life emerged with a shift in cosmologies and worldviews that was congruous with the development of modern science.

In bringing this chapter to a close, I want to expand a bit on this last point. I will throughout this book be careful in limiting my discussion of worldviews to the European perspective in many cases, because it is important to avoid assuming that ideas about science, religion, and the existence of extraterrestrial intelligence that emerged in the North Atlantic countries is representative of how other people in the past saw the universe or how they do in the present. Many authors write as though the worldview that developed Europe is somehow representative of all human civilization, but it clearly is not, even while that worldview has become increasing dominant in *contemporary* life across continents.[4]

Hut (2003: 410) has made the important point that scientific insights have been increasingly incorporated into Buddhist worldviews, and that Buddhism has had little of the trouble with Darwinian evolution experienced within Christianity, and the importance and nature of causal relationships central to physics is quite comfortable from the Buddhist worldview. That said, it remains unclear the extent to which the scientific "worldview" and the Buddhist way of life are compatible. Hut argues that science actually lacks a worldview *per se*, because it does not include propositions on the meaning of being human and concepts such as dignity and responsibility, even while scientists themselves often engage in comments based upon their work as scientists that move into these realms of human inquiry. I'm not sure I entirely agree with Hut on this point, because I think the meaning-centered ideas that scientists, and particularly many SETI scientists, frequently express have an influence on how they engage in their field of inquiry. But Hut is certainly correct that there is not the type of systematic and organized worldview operating in science that we see in institutional religions.

[4] Even while astronomy developed in many places outside of Europe, the interpretation of the stars does not necessarily lead to the assumption that there is life on other planets. Astronomy (or astrology) often is much more concerned with life here and may put "extraterrestrial" beings out in the cosmos in the form of deities (Milbrath 2010), but this has a different quality from the notion that arose in Europe and the US in the eighteenth and nineteenth centuries of civilizations similar to our own on other planets.

Chapter 2
A Brief History of Thinking About ETI

In cultural terms, the Enlightenment is the process of creative destruction with respect to medieval "godly" theocratic culture, including art, philosophy, "Christian science" exemplified by geocentric astronomy and biological creationism...

(Zafirovski 2011: 8).

The question of life on other worlds is intimately tied to how we think about the cosmos and how we conceptualize the relationship between Earth, its civilizations, and the universe. In other words, questions about the existence of alien civilizations raise questions of cosmology, which is the attempt to explain and understand the origin, structure, evolution, and ultimate fate of the universe. Cosmology is a very broad field of study, one that is pursued by scientists and theologians, although, as noted in Chap. 1, their approach normally differs significantly. Physical cosmology—which involves the work of astronomers and theoretical physicists—emphasizes a systematic examination of the structure of the universe, its history and future, and tries to identify the natural laws through which that order and structure is maintained over time. This is where we find research related to general relativity and ideas such as the Big Bang Theory. Mythological cosmology raises the same types of questions related to the history, future, and in some cases even the natural structure of the universe, but it draws upon religious texts, theological and philosophical treatises, and myths, as well as religious and spiritual experience and sometimes observation of the natural world, as a means of arriving at answers. Of course, this is where we find ideas such as the Abrahamic creation myth in Genesis or the Japanese myth of the brother/sister duo Izanagi and Izanami who were once believed to have created both the islands of Japan and many of the deities associated with Shinto.

The cosmology to which one adheres has a profound influence on the ways in which one thinks about not only this world, but the possibility of life existing on other planets. A technique I often use to engage my class on science, religion,

J. Traphagan, *Extraterrestrial Intelligence and Human Imagination*,
Space and Society, DOI 10.1007/978-3-319-10551-2_2

and the search for extraterrestrial intelligence is to ask a daily question using technology that immediately projects the results as a graph on the screen at the front of the room. Early in the semester I ask the students to respond to the statement: *The idea of extraterrestrial civilizations emerged with the advent of modern science and technology—true or false?* Most students answer that this is true, particularly if they haven't done the reading for that week. While in some ways this is an accurate response, the actual answer to this question is actually a bit vague. Democritus, writing 2,400 years ago, was aware that the Earth is round and argued that it was one of many worlds in the universe (Dick 1998); he well may have imagined the possibility of life on those other worlds. And other Greek and Roman thinkers such as Epicurus and Lucretius (ca. 99–55 B.C.) of the atomist tradition imagined a vast universe governed by natural laws that seemed likely to generate life, and perhaps intelligence, in many places. Lucretius writes in *On the Nature of the Universe* (Melville 1997: 66), that the world is the product of laws that govern the formation and structure of matter. When the elements of matter are put together in an organized way according to natural law, similar patterns should emerge in other places.

Wherefore again amid again I say you must admit
That in other places other combinations
Of matter exist such as this world of ours
Which ether holds in ardent fond embrace.
And note this too—when matter is abundant
And space is there, and nothing checks and hinders,
Then action and creation must take place.
And if there exists so great a storm of atoms
As all the years of life on earth could never number,
And if the same great force of nature stands
Ready to throw the seeds of things together
In the same way as they have here combined,
Then of necessity you must accept
That other earths exist, in other places,
With varied tribes of men and breeds of beasts.

Lucretius elegantly describes a universe governed by laws of nature in which the logical conclusion is that if the formation of matter into humans and other forms of life can happen here, it ought to be able to happen elsewhere. The atomists, of course, did not win the day, unfortunately being pushed aside in favor the Aristotelian cosmology in which the Earth sits at the center of a hierarchy of nested spheres including the Moon, planets, and stars that were fixed in place.

2.1 Narrowing Imagination

Although the geocentric worldview has its roots in Aristotelian thought, it was Ptolemy (c. AD 90–168) who developed an observational basis to support the notion that the Earth sat at the center of the universe. Ptolemy's ideas are presented in a

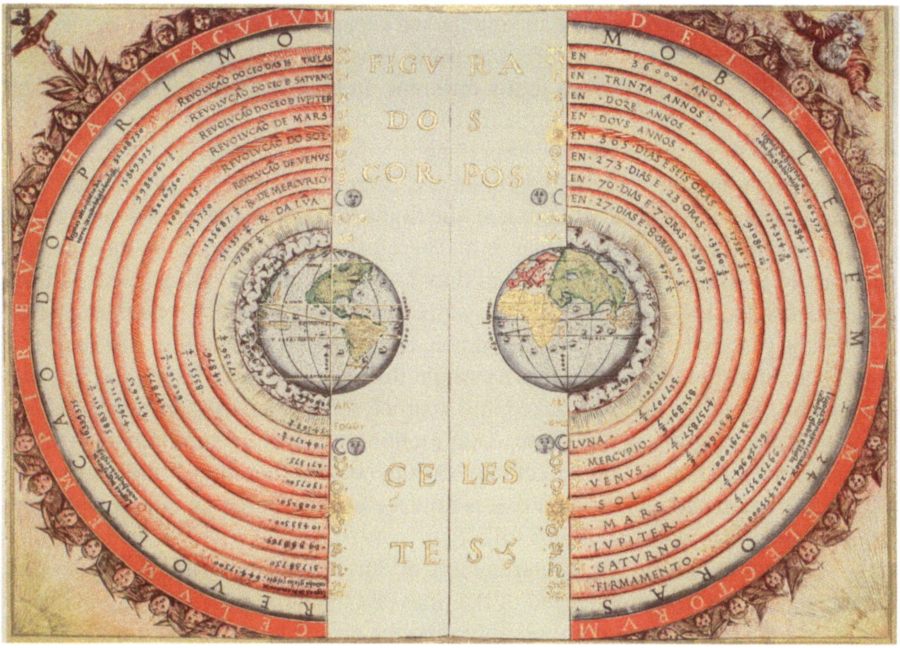

Fig. 2.1

work that has come to be known as *Almagest*, meaning the greatest, a title coined by its Arab translators. Published sometime around 150 A.D., Ptolemy's theory is complex and technically detailed, encompassing over 700 pages in 13 volumes. At the core of his ideas, however, is the creation of a system for calculating and predicting the movement of the planets. Indeed, Ptolemy's greatest contribution is not that he invented a new way of seeing the universe, but that he took ideas about the nature of the cosmos that had existed for several hundred years—again, the geocentric concept of the universe goes back at least to Aristotle—and developed a precise scientific theory that could be tied to empirical observation and used for prediction of cosmic events (DeWitt 2011: 114).

The Ptolemaic system, often simply referred to as the geocentric model, places Earth at the center of the cosmic order, where, as noted above, our planet rests amidst a nested hierarchy of celestial spheres that ranged from the Moon to the naked-eye observable planets to the fixed stars (see Fig. 2.1). In order outward from Earth, these are the Moon, Mercury, Venus, Sun, Mars, Jupiter, Saturn, and the fixed stars. This model represents observed features of the universe in terms of those things that appear to move and those that don't. Observationally, from the perspective of our planet, it appears as though the sun and the Moon revolve around us and this is the view taken by the Ptolemaic system. While this may seem counter-intuitive given our own understanding of the universe, it makes perfectly reasonable sense from an observational standpoint.

Imagine for a moment that you are sitting in an airplane traveling at 500 knots and decide to toss a ball straight up into the air and then catch it; later, after you've arrived at your destination, you sit in your parked car and try the same thing, apparently having little else better to do with your time. What's different in these two cases? Experientially, there isn't much that's different. In both instances the ball goes up and then comes down in what appears to be a straight line back into your hand. If you start to think about the two experiences, you might conclude that the ball on the plane must have actually moved in an arc, because everything on the plane was moving forward—you, the ball, the seat, the flight attendants, the annoying guy next to you, and the plane as a complete unit. Therefore, the ball had to be moving forward while it was also moving up and down, thus creating an arc-shaped motion. And because the ball and you have the same forward momentum, you both end up in the same place when the ball comes down. In contrast, while sitting in the car, you might conclude that the ball went in a straight line up and down, despite the fact that the car, you, the idiot who just backed into your left fender, and the ball are all moving along with the rotation of the Earth. That motion, however, is so minimal in terms of our experience that we can effectively treat the ball as though it simply moved in a straight line up and down—and that is how we usually perceive of tossing something up and down while standing still.

Now, suppose we change the parameters of our thought experiment slightly and assume that you don't know ahead of the time that the airplane is moving—you are flying on a cargo plane with no windows and the airline has given you incorrect or misleading information about the nature of your mode of transportation and destination (what a surprise!). What would you assume about the motion of the ball? My guess is that if you didn't already know the plane was moving and couldn't feel its motion you might assume that it was actually stationary and, thus, would conclude that the ball simply moved straight up and down. There is no necessary reason to believe that the plane is moving in this case and from an experiential perspective you, the ball, the seats, the flight attendants, the annoying guy next to you, and the ball are stationary—exactly like your experience in the car.

In fact, if you think a bit about the movement of the ball and you, it would be fairly easy to come to the conclusion that if you thought the plane were moving forward with you attached to it, then the ball would come down behind you—or hit you in the face if you didn't throw it high enough. You and the plane would move out from under the ball, which would go up and down in a straight line, while you were moving forward. Without an awareness and understanding that the entire complex of objects associated with the plane are moving forward together, it is not unreasonable to assume that you would move out from under the ball as you and the plane moved forward, while the ball went straight up and down. In other words, the context of your experience would make it perfectly logical to conclude that, despite the fact of being inside a moving object, you were actually inside a stationary object.

This is exactly what happened with many cosmologists both prior to and following Ptolemy—because experience does not *necessarily* confirm that the Earth is moving, there was no reason to work from the conclusion that it was doing so. In fact, a more reasonable and elegant solution to the problem of the motion of the Sun, Moon, and planets was to posit that they revolved around a stationary center that humans inhabited and tended to experience as a stationary object. This idea, as noted, was around long before Ptolemy. His major contribution was to provide mathematical confirmation for this intuition and to do so in a way that actually allowed for relatively accurate prediction of the movement of the Moon and visible planets. This also accounted for the fact that when observing the sky at night we do not observe a stellar parallax or the sense that the background is moving. Instead, the stars appear stationary all of the time. If the Earth were moving—whether orbiting the Sun or revolving on its axis—then we should observe stellar parallax or the apparent shift of the stellar background as our perspective changes in relation to our motion on the Earth. Because we do not experience this (due to the tremendous distances involved between Earth and the stars, making the parallax unobservable until techniques were developed to measure it in the nineteenth century), the conclusion that the Earth is stationary is supported on the basis of empirical observation from the perspective of geocentrists.

The point I want to emphasize with this example is that our basic assumption that the Earth is moving (and moving around the Sun) is a product of the knowledge generated through modern science—it is not something that we *must* naturally intuit about the world because our basic experience does not necessarily support the idea that the Earth is moving at all. This is the perspective that shaped the geocentric worldview and that dominated Western cosmology for about 1,500 years and thus provided an intellectual and cultural context within which ideas about the cosmos, and the place of humanity within that cosmos, were built and refined. The fact that it was based upon faulty observations of the natural world is irrelevant to the fact that the interpretations and conclusions based upon those observations made sense given the initial starting point. Humans are actually quite good at developing logical and rational explanations of our world that are not based upon accurate observations of our world—that's what a lot of religion is all about, but also can be a starting point in scientific inquiry. Religion has no monopoly on this tendency.

The main implications of the Ptolemaic system for our purposes here are: (1) the geocentric understanding of the universe created a cultural milieu in which Earth, and by extension humans, were perceived as inhabiting the center of the universe and (2) being at the center of everything, it became difficult to imagine a universe populated by other worlds with other intelligent beings, despite the fact that both Greek and Roman culture/cosmology allowed for this possibility.

This perspective, of course, was reinforced by the Abrahamic notion of humanity as the focal point of the creative activity of an omnipotent god who, according to the Bible, is lucky enough to be the generative core of the universe and for whom all material creatures are supposedly created (George 2005: 48). Although the Bible itself does not present a geocentric view of the universe, it does present a clearly anthropocentric worldview in that it centers itself on the relationships among humans, between humans and the rest of the created world, and between humans and the Abrahamic deity, which allows humans to see themselves as a special, and superior, element in the created order. Thus, the Ptolemaic and Abrahamic worldviews are mutually supportive or co-constructive, and within that framework the Ptolemaic cosmology provides a scientific foundation for the Christian perspective that both the Earth and humans are in some way unique in the universe.

This combination of philosophical and cultural themes shapes the basic context within which Europeans operated for about 1,500 years from the time of Ptolemy to the beginning of the Enlightenment and formed a cosmology that largely prevented the imaginations of European intellectual elites, at least, from considering the possibility of life on other worlds throughout that period. In other words, when the combined Aristotelian/Abrahamic worldview won the day, it also shut down the capacity of Europeans to imagine a universe of many worlds, inhabited by many different kinds of beings—self-centeredness always seems to have a way of narrowing one's imagination.

2.2 Expanding Imagination

The cultural and scientific innovations associated with the Enlightenment and the departure from Aristotelian/Abrahamic cosmology allowed for intellectual elites and eventually general populations to imagine a universe in which they were neither the center of creation nor alone. This expansion of imagination pivoted on the shift from a geocentric to heliocentric worldview in the sixteenth and seventeenth centuries with the emergence of modern astronomy via the mathematical and experimental approaches developed by Nicolaus Copernicus (1473–1543), Tycho Brahe (1546–1601), Johannes Kepler (1571–1630), Galileo Galilei (1561–1642), and Isaac Newton (1642–1727).

Copernicus got the ball rolling with his book *De Revolutionibus orbium coelestium* (*On the Revolutions of Heavenly Spheres*), first printed in 1543. While Copernicus maintained the Aristotelian notion of concentric spheres with the circle of fixed stars at the outer limit, he shifted from Earth to the Sun as being located at the center of European theology's celestial matryoshka doll. This allowed him to develop a simpler way of explaining the motions of the planets and, in particular, retrograde motion in which planets appear to move backwards across the sky at certain times of year. Copernicus hypothesized that he could better explain this phenomenon by placing the Sun at the center of the system of spheres and thus account for the apparent motion of the Sun and Moon as well as the apparent retrograde motion of the planets via the Earth's motion relative to the other celestial bodies. Note that Copernicus did not radically diverge from the basic paradigm of a universe consisting of concentric spheres; he simply moved the Sun to the center of that system while retaining the basic idea of spheres and also the notion that the planets moved in perfect circles. Nonetheless, the implications from a Christian theological perspective were significant because Copernicus's ideas opened the door for the possibility that humans no longer inhabited a special and unique place in the universe.

Copernicus lacked observational evidence to support his claims of a heliocentric model of the universe, but the writing was on the wall and increasingly accurate and careful observations of the skies were generating a variety of challenges to the Ptolemaic cosmology. One of these was the observation, by Tycho Brahe in 1573, of a supernova, which proved that the outer sphere of stars was not unchanging.

In addition to this, he made careful measurements to show that comets were not atmospheric and must be beyond the lunar orbit, further challenging the idea of an unchanging firmament.

The growing evidence against the geocentric model came to something of a head in the work of Johannes Kepler, a German mathematician and astronomer who in the early 1600s clearly showed that the planets moved in elliptical rather than the circular patterns assumed by Ptolemy and Copernicus. Drawing upon Tycho's observations of Mars, Kepler identified three laws of planetary motion indicating: (1) the elliptical orbits of planets, (2) that planets move faster closer to the sun, and (3) the squares of the revolution periods of the planets are proportional to the cubes of their mean distances from the Sun (Ferguson 2002). With Kepler's observations, there emerged better prediction of the movement of the planets and a much more complex understanding of how the solar system operates. These observations also further weakened the belief that Earth inhabits any kind of special position in that system.

In 1610, Galileo further undermined the geocentric model with his observations of the phases of Venus and the moons of Jupiter through the newly developed technology of the telescope. His work provided strong empirical evidence supporting the heliocentric model and Kepler's conclusions related to the motion of the planets; it also really irritated leaders in the Roman Inquisition who in 1633 tried and convicted him of heresy for supporting the heliocentric cosmology, and sentenced him to imprisonment, which was commuted to house arrest (such a kind bunch), where he remained for the rest of his life.

It was Newton who later in the 1600s put the final nail in the coffin of geocentrism when he showed that the planets were held in orbit through gravitational force. With Galileo and Newton, the Enlightenment produced the basis of empirical evidence and the theoretical framework to show that the geocentric model was simply wrong. The Earth did not sit at the center of the cosmos and, in fact, the Earth was not even particularly special in relation to other objects in the solar system and beyond. There are two points I want to emphasize here.

First, the process of generating scientific knowledge that moved us away from the geocentric model of the universe also provided an opening to think about—or given that the Greeks and Romans had already thought about it, to *rethink* about—the existence of extraterrestrial intelligence. Prior to this the capacity of Europeans to imagine other worlds with intelligent beings, or even other worlds beyond the observable realm of the planets, was very limited, because the scope of imagination was shaped by what social scientists have come to refer to as the *habitus* or overarching cultural milieu of a society that can powerfully construct the limits of imaginable thought and behavior.

Habitus is a term coined by French sociologist Pierre Bourdieu (1977, 1990), who used the concept in reference to the cognitive and social structures that motivate behavior but also shape and limit the range of ideas that naturally seem to fit into normal thought in a particular cultural context. *Habitus* is not deterministic; in other words, it does not prevent us from innovative thinking and creating new ideas, but it does have a tendency to limit the scope of our imaginations and, thus, tends to keep thought moving along a specific flow or path preventing many people within that flow from giving much consideration to alternatives. Occasionally, conditions arise that allow for rapid and dramatic changes in the course of the cultural flow, and during these periods we see significant conceptual innovations and the emergence of new ideas—this is what happened when the Aristotelian worldview won the day at the time of Ptolemy and again when that worldview collapsed and was replaced following the work of Enlightenment scientists like Kepler and Newton. *Habitus* has much in common with Kuhn's concept of paradigms of normal science, but *habitus* refers to more general cultural parameters that are embedded deeply into the minds and bodies of a group of people—so deeply that those people tend to be unaware of and unable to easily question the assumptions about worldview associated with a given *habitus*.

A good example of this can be found in the ways people in different cultures point to themselves when making a personal reference. In the US, people usually point to their chest; in Japan, people usually point to their nose. This example may seem a bit trivial, but it indicates the depth of *habitus*. In neither context do people give much of any thought to where on their body it is natural to point when referring to themselves. In fact, they pick this up through mimetic processes of cultural learning—as children, they see others do this and simply copy what they see and often do so unconsciously. When the situation arises to point to oneself, one just naturally points to nose or chest, depending upon the cultural context in which one was raised, without giving any thought to where is the proper place to point—it is completely internalized. This is what Bourdieu means when he talks about *habitus* and this can apply to trivial actions like self-pointing or to cosmologies like geocentrism or heliocentrism. Again, this is not cemented into our bodies and psyches; one can change, either consciously or unconsciously as one interacts with others in a given cultural environment or with outsiders who challenge conventional ideas. When I spend long periods of time in Japan, I find myself bowing on a regular basis without thinking about it simply because I pick it up from people I see around me. And I often continue to do this when I get back to the US for a while, until I unlearn—or un-embody—the practice.

One important difference between *habitus* and a paradigm is that unlike the subculture of science, human society more generally does not necessarily have a built-in assumption that inquisitiveness and openness to ignorance is to be valued. Thus, there often is little or no incentive to develop ideas and practices that run counter to general patterns of thought and behavior. In fact, most, perhaps all, societies tend much more in the direction of encouraging tacit conformity and provide few contexts in which individuals or groups can challenge the accepted norms of thought and behavior.

Second, although scientific paradigms may have more openness to radical innovation than the paradigmatic cultural patterns associated with *habitus*, a scientific paradigm is, in many respects, a subset of *habitus*. It is a way of seeing the world that shapes the parameters of acceptable scientific inquiry and deeply influences the ways in which most scientists conduct their research. Like with the broader culture, there are gatekeepers who shape and limit the range of acceptable questions and interpretations—think about the peer review process, as noted earlier—but there also are the more deeply embodied assumptions that have an influence on our capacity to imagine new questions and develop new interpretations. The Ptolemaic cosmology was based upon observation, but those observations were, in fact, wrong. Nonetheless, in part because they were based upon observations, they came to represent and construct reality and the overall *habitus* in which the Ptolemaic system emerged and was elaborated upon for 1.5 millennia, representing a cultural context in which challenging either the basic assumptions and calculations of Ptolemy or the theoretical framework of geocentrism was difficult and for much of the period very nearly impossible.

It would be a mistake to think that modern science operates any less within the context of a *habitus* that influences the limits of human imagination. The main difference is that the current cultural parameters are much more open to innovation and change than those of, say, the tenth century, at least when it comes to physical cosmology. However, scientists continue to have their questions and their conclusions shaped within the cultural context in which they conduct their research. One only need think about the recent debates over stem cell research, or creationism vs. evolutionary theory, to see that broader cultural trends and themes have the potential to significantly influence the course of scientific research.

We will return to this point later in the book, as it is important for understanding the manner in which SETI research has developed over the past 60 years. For now, I want to return to the historical dimensions of SETI, because the story does not end with Newton—in fact, the emergence of Newton's ideas are really just the beginning. His realization that there was a consistency to how the universe operated—a mathematics through which the motion of objects in the universe could be described and predicted—further opened a door to thinking about the existence of extraterrestrial intelligence. Newton showed that the laws of motion he described mathematically not only applied to the motion of objects on Earth, but also to celestial bodies. There is a very logical and ultimately inescapable conclusion that arises here: *If the universe is governed by universal laws, then life may not be limited to Earth.*

This conclusion was not missed by other thinkers of the time and by the late eighteenth century intellectuals such as Thomas Paine were postulating the possibility of numerous worlds beyond Earth. Paine wrote in his three-part pamphlet, *The Age of Reason* (1890: 66), published between 1794 and 1807, that "to believe that God created a plurality of worlds at least as numerous as what we call stars, renders the Christian system of faith at once little and ridiculous and scatters it in the mind like feathers in the air. The two beliefs cannot be held together in the same mind…" Indeed, Paine developed his notions about a plurality of worlds, as well as his critique of Christianity, around the idea that there is a basic consistency to the laws of the universe that should be open and visible to all beings, whichever planet they might inhabit: "The inhabitants of each of the worlds of which our system is composed, enjoy the same opportunities of knowledge as we do. They behold the revolutionary motions of our earth, as we behold theirs. All the planets revolve in sight of each other; and, therefore, the same universal school of science presents itself to all" (Paine 1890: 72).

This is a significant point that became amplified in the nineteenth century with the awareness that not only the laws governing physical motions of inanimate objects, but also laws governing biological change, might be universal. Right at the beginning of the century, there emerged a realization that the biological world might be subject to uniform laws that shaped its development and the process of change, just like what had been found by early physicists interested in the movement of celestial and other objects. French naturalist Jean-Baptiste Pierre Antoine de Monet, Chevalier de Lamarck, or Lamarck for short, published research in which he explored two basic themes: (1) that variation in animals is the product of the environments in which they live and (2) that life developed through specific forces that generated order and structure to organisms. The importance of Lamarck was that he was the first to develop a coherent theory of organic evolution, although he lacked an understanding of the process of inheritance and, thus, his ideas were eventually displaced by the much more elegant and accurate work of Charles Darwin and eventually Gregor Mendel. But Lamarck's fundamental observation that the development of life was a result of observable natural laws and that biological organisms evolved not at the whim of a god, but in relation to their environment, raised the possibility for thinking about all life as being governed by observable natural laws.

Evolution as a way of seeing the biological world became an increasingly powerful tool, of course, with the publication of Darwin's *Origin of Species*. This book, which is among the most important ever written, not only changed our understanding of the natural world, it also challenged our understanding of ourselves, because Darwin's process of natural selection effectively eliminates the necessity of a divine creator who has intimate interest in the design and development of life, either human or non-human, and who for some strange reason decides to put it only on Earth. Natural selection responds to the apparent improbability of a world of complexity arising either as a result of chance or some activity of an intelligent designer by recognizing the enormous power of accumulative change over a very long period of time (Dawkins 2006: 147).

Darwin was freed from the biblically inspired silliness of an absurdly short time span to geological history by the research of Sir Charles Lyell (1797–1875). The Earth's surface, Lyell realized, was formed over vastly long periods of time through uniform processes that continue to operate. This contrasted starkly with the widely accepted notion in England at the time (and elsewhere in the European sphere of influence, as well as, sadly, the contemporary US) of changes in the Earth occurring as a result of catastrophic events such as Noah's flood described in the Bible. Lyell's work was important for at least two key reasons. First, it was based upon empirical observations of the world, not upon culturally idiosyncratic myths such as those found in the Bible, Quran, or other religious texts. Second, it provided the long geological time span necessary for Darwin to show how minute changes in organisms can have sufficient time to accumulate (Darwin 1859: 282), forming the types of complex structures we see in humans and other animals in our world. Indeed, Darwin builds his theory of natural selection within the confines of an understanding of the world that necessarily has extremely long periods of geological time allowing for the possibility of the emergence of considerable biological complexity and variation. Given the incredibly long time that Earth has existed (by human standards), it becomes quite imaginable that the accumulation of minor changes in organisms would lead to complex structures like chimpanzees, dogs, and *homo sapiens*. Egotistical deities who create humans, and everything else, for their own warped pleasure and desire to be worshiped are no longer necessary in this way of seeing the natural world.

This rather brief history of the development of modern physical and biological science brings us to one of the key points of this chapter—that the emergence of modern science changed the capacity of humans to imagine the possibility of life on other worlds. First, this change was stimulated by the revolutionary work of astronomers from Copernicus to Newton that allowed humans to imagine a physical universe in which the Earth was no longer at the center or even particularly important. Second, the equally revolutionary work of scientists such as Lyell and Darwin showed that geological and biological change also can be understood in terms of observable laws, or at least observable patterns, and that it is completely reasonable, given the long periods of time involved, for minor changes to accumulate into the implausible forms of pigs, dogs, snakes, or even theologians. If the laws of physics and biology are uniform, then there is no reason to think life, or even intelligent life, could not emerge on other planets, and perhaps on many other planets.

By the middle of the nineteenth century, the universe humans inhabit—well really Europeans and their colonial descendants, since many in Asia such as Buddhists and Taoists never had this sort of worldview—had shifted from being a tiny place with a short history centered on a special Earth created by an egotistical god, to a vast place with a long history in which Earth was likely a very minor player and the reason for and source of creation was open to a great deal of uncertainty and debate. In short, from the beginning of the Enlightenment to the middle of the nineteenth century, human minds were not only opened to the way the universe actually is, they also were opened to imagine how it might be given what we were increasingly coming to know about how it is.

2.3 Imagining Aliens

By the later part of the nineteenth century, the stage had been set for the ability to imagine life on other worlds and several technological developments helped in energizing both scientific and popular curiosity related to the possibility of extraterrestrial civilizations. Perhaps the most notable of these improved technologies was the refinement of telescopes giving astronomers the capacity to view other planets in our solar system with increased clarity. And one of the best targets for observations was found in the planet that came to be viewed as analogous to Earth—Mars.

Lane has argued convincingly that the creation of Mars as a terrestrial analog drew heavily from the emphasis on geography and map-making that had become popular in the later nineteenth century, or as Lane (2011: 18) puts it "disciplinary geography and its imperial influences were fundamental to the emergence, entrenchment, and duration of the inhabited-Mars hypothesis." The process of mapping Mars, which occurred at a variety of different telescopes around the world, was, according to Lane, one of imprinting a terrestrial image on the Martian landscape through the creation of drawings generated through careful, albeit flawed, observations of the Red Planet.

This process took flight in 1877, when Italian astronomer Giovanni Schiaparelli trained his telescope on Mars and observed a network of seemingly straight lines he called *canali*, which actually means channels in Italian, but which was translated into English as canals: an unfortunate choice, because in English the word implies something constructed, rather than simply a natural channel. An intellectual debate rapidly emerged both in scholarly journals and in public media about the nature of Mars and the likelihood that there existed a civilization on the planet.

Indeed, it appeared to some, from the low resolution images that could be obtained by astronomers of the time, that Mars was a world with running water, oceans, and long channels along which the water moved. Both professional and amateur scientists began to argue in favor of the idea that Mars harbored intelligent life and that improved technologies would eventually confirm the idea that features such as the canals were artificial in nature (Lane 2011: 1). Some of these writers argued, as did one anonymous contributor to the British popular magazine *Chamber's*, that "the Martialites are probably much further advanced in the arts and sciences" than humans (quoted in Lane 2011: 1). In the US, it was Percival Lowell, a wealthy Boston businessman with an interest in astronomy, who at the end of the nineteenth century became fascinated with Mars and emerged as the leading proponent for the theory that Schiaparelli's observations were not simply the products of natural movement of water across the surface of the planet, but were canals constructed by an advanced Martian civilization (Hoyt 1996).

As Dick (1998: 31) has noted, Lowell did not develop his ideas about Martian civilization in a vacuum. His theory emerged within a scientific milieu where there was debate about the nature of the markings on the surface of Mars observed by Schiaparelli and others, and there were questions about the extent to which the existence of canals had or had not been empirically confirmed by the observations of other astronomers. Some astronomers reported canals, while others only saw shaded areas on the surface of the planet (Dick 1998: 28). Indeed, when Lowell weighed in on the scientific discourse about canals on Mars, most scientists believed that the canals were cracks in the crust of the planet (Dick 1998: 31).

In a series of lectures delivered at the Lowell Institute in 1906 and then published in *Century Magazine* and eventually as a book, Lowell argued a case for the presence of water, vegetation, and intelligent life on Mars that drew upon a variety of scientific frameworks ranging from astronomy to geology to Darwinian evolution. The gist of Lowell's argument was that the physical features of the planet—the "straight" lines and "oases" he was convinced he had observed through his telescope in Flagstaff, Arizona—could not be explained as natural phenomena (Lowell 1908: 196). Indeed, from Lowell's perspective, the circular structure of the oases, the straightness of the canals, their relations to each other, and a set of mathematical calculations about the movement of water, could only suggest one thing. "The deduction is inevitable; [the water] must have been artificially conducted over the surface of the planet. We are left no alternative but to suppose it intelligently carried to its end" (Lowell 1908: 202).

Lowell was a convincing writer and powerful orator who could make his case in such a way that had many scientists and far more lay people agreeing that the evidence for life on Mars was becoming increasingly conclusive. Although there were many detractors, it was not until the emergence of photographic images of Mars taken through the ongoing development of larger and higher resolution telescopes that it became clear that Lowell was wrong and there were no obvious signs of intelligent life on our celestial neighbor. In fact, by 1907, Lowell was backpedaling as improved techniques in spectroscopy and photography not only provided much better data, but also pointed out the inherent subjectivity of the astronomical maps that had been generated up to that point (Lane 2011: 57). Although by the early part of the twentieth century the idea of canals on Mars had largely died out among scientists, among the public the notion continued to persist—indeed, growing up in the 1960s I remember reading about and believing that there were canals on Mars and people still talked about canals as physical features of the planet. And despite lack of scientific evidence to support it, the idea of intelligent life and even a great civilization on Mars continued to fascinate both some scientists and the public into the early 1960s.

2.4 It Came From Outer Space

Of course, scientists were not the only ones imaging alien civilizations around the beginning of the twentieth century. Parallel to the growth of astronomy in the nineteenth century was an emergence of a new genre of literature—science fiction. In 1865, Jules Verne published *From the Earth to the Moon*, in which he imagines members of the Baltimore Gun Club in the post Civil War US building a huge cannon to send a spaceship carrying three people to the moon. Perhaps the most notable of the era's science fiction authors was H. G. Wells who, picking up on the Mars fever of the times, published in 1898 *The War of the Worlds* about an invasion from Mars and then in 1901 published his book *The First Men in the Moon* in which the protagonists discover an advanced civilization of insect-like creatures crawling around our nearest neighbor.

It is *The War of the Worlds* that I think is most instructive in exploring how the human imagination had opened to the idea of extraterrestrials. Not so much in the publication of the book, but in Orson Welles' broadcast in 1938 of the story as a radio drama for *The Mercury Theater on the Air* we see the extent to which the broad cultural conceptualization of alien intelligence had changed from less than 100 years earlier. Despite the fact that by 1938 the idea of Martian canals had been debunked, many in the public continued to believe that the canals existed and represented evidence of a Martian civilization. Welles, of course, took some liberties for dramatic purposes and presented the broadcast as though it was a news report of an actual event, although if one had tuned in from the beginning it was clear that it was a dramatization. The broadcast—in part due to general war fears that were becoming increasingly widespread by 1938—sent a scare across parts of the US and, at least at some level, a mild panic ensued about the possibility that Martians had, in fact, actually landed. Many police stations received calls asking about the verity of the reports as did radio stations and newspapers.

Orson Welles' broadcast was, of course, only one example of the expansion of science fiction stories in the public media. Comic strips like Buck Rogers (first appearing in 1928) and Flash Gordon (first published in 1934) brought the idea of space travel and exploration to the general public and represent expressions of a growing capacity to imagine the idea of intelligent life on other worlds and the development of technologies that both might get us to those alien worlds or bring the aliens to Earth. Superman, which first appeared in DC Comics in 1938 and became an American icon, is particularly interesting because he represents a human-like creature with unique—and superior—abilities who comes from an alien civilization, albeit a dying one, that is technologically, intellectually and morally superior; a fact that seems to be underscored by the inability of humans who see him constantly at the Daily Planet to notice that Superman is Clark Kent sans eyewear. In Superman, we see the development of the idea that aliens might be able to use their superior abilities to help humans emerge from their varied and complex problems and tendencies toward self-destruction and violence.

These fictional characters are the product of an early to mid twentieth century society in which new technologies were emerging very rapidly. Radio, the airplane, the automobile, motion pictures, Goddard's liquid fueled rocket launched in 1926, etc. provided the foundation for creating fictional characters from either an advanced human future or from other worlds. In other words, the shift from a geocentric to heliocentric worldview had, by the end of the nineteenth and middle of the twentieth centuries, opened the door for humans to imagine varieties of intelligent life on other planets, as well as to contemplate a future in which humans travelled to those other planets. Science fiction becomes a vehicle to think not only about aliens, but also to think about ourselves as we construct our own society in contrast to either superior/altruistic or dangerous/imperialist alien intelligence as well as in contrast to an imagined future.

Anthropologist Arjun Appadurai (1996: 31) argues that with the growth of global economies and the expansion of generative processes of cultural creation has come the emergence of "the imagination as a social practice." By this, he means that the capacity to engage in imaginative practices is no longer an elite pastime or the idle contemplation of the scholar, but has become "an organized field of social practices, a form of work (in the sense of both labor and culturally organized practice), and a form of negotiation between sites and agency (individuals) and globally defined fields of possibility." From Appadurai's perspective, the capacity to imagine has become central to human agency and in the process of that development there has formed an imaginary or a "constructed landscape of collective aspirations." Appadurai is interested in the emergence of this imaginary in relation to the expression of political power through media and other forms of globalizing structures and practices and the creation of what he calls ethnoscapes, or landscapes of people who create the shifting world around us and often briefly move in and out of our lives in a globalized world. The emphasis on this idea is one of instability and change and social practices that bend and warp more stable communities such as kinship and friendship networks.

I tend to disagree with Appadurai's notion that the imagination as a social practice is something new; human imagination is always a kind of social practice, even if the capacity to imagine is severely limited by a given social and ideological context. Humans always live within the context of an imaginary—or a way of putting together reality that links how we think the world is with how we think it ought to be. In other words, reality is not something out there that we touch, it is a consequence of the interaction between the physical and social context and individual imaginations that generate feelings among individuals with whom they share seemingly common ideas and experiences. It is in those feelings of a shared reality that we find imaginaries, which continually change in relation to the ways in which individuals interact with and imagine their surroundings.

What we see in the early part of the twentieth century in the US and Europe, at least, is the manifestation of a new type of imaginary with previously unimaginable repertoires of technologies, images, and narratives that stimulated a way of imagining humanity and its relationship to the cosmos that was only a few hundred years earlier largely unimaginable (Appadurai 1996: 35). In this imaginary, not only are people, with their various values and ideas, moving around the world and coming into contact with each other at an increasingly rapid pace, but the "world" as both a social and geographical construct is no longer limited to our planet. The conceptual geography of the Earth shrank with new technologies like radio and air travel, and along with it the conceptual geography of the universe shrank as it became possible to imagine a universe in which alien intelligences might travel to Earth and, eventually, we might travel to their worlds as well.

World War II, of course, accelerated the pace of technological innovation via the creation of jet airplanes, V1 and V2 rockets, and nuclear bombs, among many less visible innovations generated out of necessity during the conflict. My aim here is not to run through a history of technological innovation during the first half of the twentieth century; rather, I am interesting in stressing the idea that by the end of World War II, Americans, in particular, had experienced a array of technological and scientific advances combined with imaginative representations in fiction of a potentially widely inhabited universe such that the average person was faced directly with the idea that humans might be only tiny members of a cosmic chorus instead of a lone tenor belting out *Fly Me to the Moon* in front of the bathroom mirror.

For Americans, the end of World War II brought many significant changes particularly related to geopolitics. Our military was spread out across the world and, at least for a brief interval until the Soviet Union emerged as a nuclear foe, our political clout was uncontested. The end of the war also brought an influx of new ideas and people that came from conquered lands. For our purposes here, the most important of these arrived in the form of a German scientist named Wernher von Braun (1878–1972), who had been a key figure in the German rocket program during the war and became the central scientist in the creation of the space program in the US.

One of von Braun's biographers, Michael Neufeld (2007: 223), describes the German engineer who was brought to the US at the end of the war as "a true believer in science and technology, with a strong tendency to scientific utopianism." Indeed, von Braun became much more than a rocket scientist. As his role in the American rocket program expanded in the 1950s, von Braun became the chief salesman of a vision of the future in which humans colonized space and traveled to the stars. Where space travel had been limited to the realm of Flash Gordon and Buck Rogers during the late 1920s and 1930s, in a series of articles for *Collier's* in the 1950s, von Braun made a case that humans were ready to transfer space travel from science fiction to science fact. When the first issue of *Collier's* focused on space travel appeared in 1952, von Braun became a public spokesman for the space age, appearing on several television shows with major media characters like Dave Garroway and Garry Moore (Piszkiewicz 1998: 73). As *Collier's*—and von Braun—imagined the future of humanity, it was one in which humans engaged in a "conquest of space" drawing upon another imaginary related to American society—the frontier. The centerpiece of the *Collier's* discussion of space was an article by von Braun called "Crossing the Last Frontier," in which the author described a time 10–15 years in his future when humans had built a huge space station orbiting 1,075 miles above the Earth and built in a donut shape that would be 250 ft across. From this space station, von Braun describes a trip to the Moon as "just a step" and clearly imagines a rapidly coming future in which humans routinely access space.[1]

Obviously, von Braun's predictions were a tad optimistic, but they helped to extend the frontier imaginary of the American West to include the realm of space and in so doing further opened the imaginations of Americans (and others) to the possibility of extraterrestrial intelligence. If humans could go into space, why not intelligent beings from other planets? The cultural milieu of the 1950s, in fact, had many elements that contributed to the expanding of the American imagination to include the idea that humans might not be alone in the universe. Numerous movies—mostly bad ones and a few good ones—were produced by Hollywood depicting a wide array of potential encounters with aliens, from the morally and technologically advanced Klaatu in *The Day the Earth Stood Still* trying to help the infantile humans grow up, to the surreptitious pod people invading the Earth by taking over human bodies in *Invasion of the Body Snatchers*. On the silver screen, the imaginary of space travel and alien beings from space was playing out initially in black and white and then in full Technicolor.

2.5 It Came from Earth, Too

I have briefly explored a very long period of human history to make a simple point: By the 1950s, humans—and particularly Americans—had taken the seeds of a new imaginary that emerged during the Enlightenment and grown them to the point that it was now fully possible to conceive of the idea that humans were not alone in the

[1] It's interesting to note that as I write this page, NASA has just announced that the Voyager space probe has moved into interstellar space—the first human-made object to depart the solar system.

universe and to even imagine what alien others might look like, with their quasi-humanoid vast craniums, bug eyes, and the like. Starting with the Copernican revolution and the realization that our planet was not at the center of anything, an interpenetrating flow of new ideas and innovative technologies combined to lead humans, by the end of the nineteenth century, to a point where they could fairly easily imagine other planets inhabited by intelligent species. And by the 1950s, this imaginary had intensified and broadened significantly as it became clear that we were on the verge of developing the technologies that would allow humans to travel into space. People now had the cultural tools to imagine aliens—they could think about what they might look like and could ponder other planets where civilizations, and ones likely to be more capable than our own adolescent or toddler one on Earth, built great machines like the 30 km^2 underground computer of the Krell and Robbie the Robot in the movie *Forbidden Planet*. It was in this cultural milieu that scientists in the 1950s began to ponder the possibility of designing research projects that might generate the empirical evidence needed to determine if extraterrestrial intelligence actually existed.

Chapter 3
Are We Alone? The Emergence of SETI

Our sun is one of 100 billion stars in our galaxy. Our galaxy is one of billions of galaxies populating the universe. It would be the height of presumption to think we are the only living things in this enormous immensity.

— Wernher von Braun, quoted in the New York Times, 29 April 1960

At this point in our exploration, it should be fairly clear that I see the rise of SETI as a product of the confluence of cultural and technological innovations that developed from the Enlightenment onward and generated in Europeans, Americans, and eventually many others, the capacity to imagine a universe populated by non-human intelligent species with whom we might be able to communicate. We were not, as noted earlier in the book, the first to think about such things. The Greeks imagined a universe in which it seemed likely that other civilizations could exist and the follower of Democritus, Metrodorus of Chios, in the fourth century BCE, was able to write of an infinite universe populated with other species: "To consider the Earth as the only populated world in infinite space is as absurd as to assert that in an entire field sown with millet, only one grain will grow" (in Kargel 2004: 4). But among the Greeks, these ideas were the imaginings of the elites—the educated and literate—and eventually the Ptolemaic/Christian worldview derailed their insights in such a way that it generated a different kind of imaginary in which the idea of alien others was difficult to conceive. It was not until the invention of technologies such as the telescope, allowing for better collection of empirical data, and the creation of calculus as a means of mathematically describing the universe through the work of Descartes, Newton, and Leibniz that cracks in the geocentric imaginary began to become gaping holes opening a path to new ways of seeing the universe and our place in it. By the middle of the nineteenth century technologies of mass literacy and education—newspapers, magazines, the telegraph—allowed for the emergence

J. Traphagan, *Extraterrestrial Intelligence and Human Imagination*,
Space and Society, DOI 10.1007/978-3-319-10551-2_3

of a broad imaginary in which the general public could think about and debate the existence of intelligent extraterrestrials. And in the twentieth century this intensified with the invention of radio and television and much later the Internet. Without that confluence of technological and cultural innovation—and the corresponding generation of a new imaginary—the scientific search for extraterrestrial intelligence that began in the second half of the nineteenth century with astronomers like Percival Lowell and continues to the present would not have been possible.

As discussed in the previous chapter, the ideas of Lowell (and others who followed him) about Mars were biased in many ways, perhaps most importantly by a personal desire to observe a civilization resident on our cosmic neighbor, and this deeply influenced his interpretation of the data he collected. Indeed, the wish for the existence of an alien civilization just around the corner powerfully influenced the manner in which data were both collected and interpreted. By the 1920s, it had become clear that the empirical evidence for canals and a complex water-based civilization on Mars was weak at best, even if there remained a viable hypothesis that Martian vegetation might exist. From the 1940s onward even the idea of vegetative life had become questioned and the notion that animal life might exist was deemed highly improbable due to the scarcity of oxygen in the atmosphere (Dick 1998: 48–49). Nonetheless, throughout the 1950s there remained active interest in the possibility of vegetative life on Mars among scientists, even while there were ongoing differences in opinions about the manner in which data related to the presence of water and the nature of the atmosphere should be understood. And the public discourse on Mars continued to raise the possibility that there might exist in the present or the distant past a great civilization on Mars—indeed, there continue to be those in the early twenty-first century who think that certain natural, geological features of the planet are actually evidence of a past civilization capable of large-scale monumental architecture.

As Dick (1998: 51) has pointed out, debates within the scientific community related to the possibility of Martian vegetation pointed to differences in both methodological approaches and worldviews among scientists, "with one extreme group much more likely to go out on a limb and to extrapolate than the other." Some astronomers were content with presenting data and avoiding interpretive turns in which they attempted to address the question of extraterrestrial life, while others gathered data with the aim of answering said question. While these debates continued, two important technologies were becoming powerful tools to begin addressing the question of both extraterrestrial life and extraterrestrial intelligence—space travel and radio astronomy.

3.1 Leaving Earth

In the postwar milieu of the US, new technologies were not only supporting the emergence of an imaginary in which aliens seemed both possible and even a potential threat (think of the UFO sightings that became common after World War II).

Technologies like television were also contributing to the emergence of another imaginary that involved an alien other—the scare over communism exemplified in the tactics of Eugene McCarthy (Doherty 2013). Fear and paranoia about the dangers of communism as well as the dangers of atomic power in the 1950s were evident not only in the news media, but also many popular science fiction movies.

As Hendershot (1999: 7) has noted, the prehistoric monsters, body-invading pod people, Martians, giant ants, and intelligent vegetative "things" from another world displayed to audiences in movie theaters across the nation were part of a public discourse of paranoia that expressed insecurity about life in the face of both the Soviet threat and the potential for atomic disaster. Extraterrestrial aliens were one way to represent fears about communism and atomic annihilation, as well as to convey the possibility that we might be saved from our stupidity by an alien civilization that had matured past these political problems, as is evident in the messianic figures of characters like Klaatu from *The Day the Earth Stood Still* (Hendershot 1999: 23), although most 1950s aliens were more of the sinister variety seen in movies like *Killers from Space* in which the nasty aliens wanted to eradicate humanity using giant bugs and reptiles created from our nuclear testing. And then there is that masterpiece of awfulness, 1953s *Robot Monster*, in which Ro-Man Extension XJ-2 has exterminated all but eight of the humans on Earth with his Calcinator Death Ray—apparently even evil robot monsters from space wearing gorilla suits and helmets on their heads have technical limitations, but I suppose 2,665,865,384 out of 2,665,865,392 is a pretty good percentage (99.999999699 %).

The pivotal point in the experience of this imaginary came in the late 1950s with the launch of Sputnik by the Soviet Union on 4 October 1957. The Vanguard program, which was the American effort to launch a satellite into orbit, was moving along—albeit with many problems related to the performance and design of its rocket—and the American public was generally under the assumption that their country would be the one to inaugurate the space age (Dickson 2001: 11). Americans were stunned by the fact that the Soviets had beaten them to the punch line and shown clear evidence of technological superiority when it came to spacecraft, a point summed up in the cynical comment of Wernher von Braun, who lamented, "we will have to pass Russian customs when we finally reach the moon" (Piszkiewicz 1998: 103).

Others have discussed the history in detail, so I will avoid that here. Rather, I want to emphasize the fact that in the late 1950s it became reasonable to think humans would eventually leave the planet and perhaps even do so in large numbers in the not too distant future.[1] The launch of Sputnik, as well as earlier technological achievements such as breaking the sound barrier in 1947 by Chuck Yeager in the Bell X-1, provided evidence in the public discourse that the stars might be within reach of

[1] This is not to say that everyone thought humans would, or should, travel to space. Bob Ward (2005: 88–89), in his biography of Wernher von Braun, notes that many Americans thought that it was either impossible to go to the Moon or that it was somehow contrary to the Bible and, thus, against the wishes of the Christian god for humans to depart the planet. And von Braun himself was occasionally accused of being a nutcase due to his insistence on the idea that humans would travel into space.

humanity. It is striking when one watches movies from this era the extent to which Americans, at least, were unaware of the truly daunting challenges involved with space travel—rarely are the kinds of distances involved in travel to other planets in the solar system, let alone other stars, accurately represented, usually being vastly under-estimated, with aliens arriving at earth from planets 10, 20, or 100 million miles away, but clearly not from our solar system. However, despite the lack of accuracy, the combination of science fiction stories (both in print and film) with the rapid flow of technological breakthroughs related to space travel strengthened the idea of extraterrestrial civilizations existing somewhere in the universe, as well as the possibility that some of those aliens might even be visiting Earth. If our toddler steps into the cosmos were able to happen in such a short time—remember the Wright Brothers had flown only 50 years earlier—then what was there to prevent a much older alien civilization from faring into space and even visiting our planet?

While rocket scientists were pursuing new technologies that captured the imagination of the public, a few other scientists were also thinking about how a different technology—radio astronomy—might be used to determine if extraterrestrial civilizations actually existed. In general among the scientific community in the 1950s, the search for extraterrestrial intelligence was not taken particularly seriously, but there had been a few attempts earlier in the century to identify radio signals from an alien civilization. From 1919 to 1922 popular publications such as *The New York Times* and *Scientific American*—both well-respected sources for information on science—had covered claims that none other than inventor of the radio Guglielmo Marconi had received evidence of radio transmissions from Mars (Brown 2005: 18). These claims were part of a public discourse not only about the existence of Martian civilization, but about the possibilities that it might be possible to communicate with the Martians by using the recently invented technology of radio telegraphy.

In the end, Marconi's claims were discounted and little interest in the possibility of using radio-based technologies was evident until the late 1950s when Philip Morrison and Giuseppe Cocconi, both astronomers at Cornell University, published a paper in *Nature* (19 September 1959), in which they raised the idea of searching the microwave spectrum for signals from extraterrestrial civilizations. This article is considered a cornerstone publication in the scientific search for extraterrestrial intelligence. Morrison and Cocconi begin with a general summary of the understanding about life on other planets as of 1959:

> No theories yet exist which enable a reliable estimate of the probabilities of (1) planet formation; (2) origin of life; (3) evolution of societies possessing advanced scientific capabilities. In the absence of such theories, our environment suggests that stars…can possess planets, that of a small set of such planets two (Earth and very probably Mars) support life, that life on one such planet includes a society recently capable of considerable scientific investigation.

Interestingly, Morrison and Cocconi take the strong position that there is "very probably" life on Mars and use that as a basis for assuming that our (empirical) environment suggests that life should exist elsewhere, since it appeared to have happened twice right here in our own stellar neighborhood. This is one of many important assumptions that characterize the article and that have continued to shape SETI until the present.

For example, Morrison and Cocconi argue that some of the societies on other planets should be very old, perhaps even old in a sense "comparable with geological time." Furthermore, working from the assumption that the universe should have an abundance of life, "[i]t follows," according to Morrison and Cocconi, "that near some star rather like the Sun there are civilizations with scientific interests and with technical possibilities much greater than those now available to us" and to the beings who populate such civilizations our own star would appear as the "likely site for the evolution of a new society" (Morrison and Cocconi 1959: 844).

From these initial assumptions, which are in fact entirely speculative, the authors detail some specifics about frequencies, bandwidth, and potential target stars from which a radio signal might be received. As Shuch notes, brief as it is, this article became a blueprint for SETI research that continues to have influence into the present (2011: 5). But these assumptions are not without problems.

Obviously, the fact that the likelihood of Martian life has dwindled somewhat with improved methods for observing the planet weakens the argument for abundance of life being grounded in any empirical evidence from our cosmic neighborhood—Earth remains the only place where we have found inconclusive evidence of life. Perhaps more important for my argument here is the assumption that those alien civilizations should be like us and thus that they might recognize our star as a likely home to a civilization. At this point in time, we have no reason to believe that ours is the only type of star that would generate a society capable of radio transmissions and, in fact, there are a few exoplanets in habitable zones that orbit stars different from our own, such as Kepler-186f, which has a radius similar to Earth's but orbits the red dwarf Kepler-186 about 500 light-years from Earth. Nor is it necessary to assume that a civilization with which we might come into contact would be much older than our own, although obviously the longer a civilization has been broadcasting, the more likely we are to intercept a broadcast.

The problem we face is that all we have is one planet, one fragmented "civilization," and absolutely no concrete evidence for life anywhere else. Thus, we have no basis for comparison and no means by which to draw even tentative conclusions about the nature of other civilizations on other planets orbiting other stars or to even guess at the likelihood of their presence—anything we say that is in any way predictive about the nature of life on other worlds is conjecture.

This also leads us to a significant issue in much of the writing that has been done in SETI on the possible nature of alien "civilizations." That is the tendency to anthropomorphize institutions and other forms of social organization as *things* with "scientific interests" rather than to recognize that our own "civilization" on Earth is, in fact, extremely fragmented and quite difficult to assign any interest to other than self-preservation and that is even open to question. Alien civilizations are somehow imagined as being homogeneous and displaying scientific interests and general unity, despite the fact that we have absolutely no evidence from Earth that this is a likely pattern of social organization among unambiguously intelligent species.

Specious assumptions aside, Morrison and Cocconi end with an important comment that has become something of the mantra of SETI researchers when confronted with the fact that several decades of searching has produced nothing in the way of positive results. Morrison and Cocconi argue that while it is difficult to estimate the probability of successfully finding an alien radio signal, "if we never search the chance of success is zero" (Morrison and Cocconi 1959: 846). This key point has buoyed the spirits of astronomers engaged in SETI in the face of low odds of success and a long-term inability to produce any empirical results supporting the idea that extraterrestrial intelligence exists. The idea also has been widely used as a basis of justification for continued investment of both time and money in SETI. Let me be clear that I don't point out the assumptions in Morrison and Cocconi's article to devalue their work—it is important and set forth an entirely new area of scientific research. But all forms of research start with assumptions and some of the assumptions that have continued to shape SETI are clearly evident in this germinal publication.

3.2 ETI in the Scientific Imagination: The Drake Equation

Around the same time that Morrison and Cocconi were thinking about radio astronomy as a means to seek evidence of extraterrestrial civilizations, another astronomer, Frank Drake, recognized the potential of the technologies of radio astronomy for revealing the existence of alien civilizations. If there were civilizations out there that could send out radio signals, we had the technology in radio astronomy to potentially intercept those signals. Drake initiated the first observational study designed to detect signals from extraterrestrial intelligence in 1960 with his Project Ozma, named after the princess of Oz in the L. Frank Baum books. After completing his PhD in astronomy at Harvard, Drake took a position at the National Radio Astronomy Observatory at Green Bank, West Virginia, where he was head of the Telescope Operations & Scientific Services Division and conducted planetary research and cosmic radio source studies. Drake used the National Radio Astronomy Observatory to look at two nearby stars, Tau Ceti and Epsilon Eridani, focusing on a part of the radio frequency spectrum near the 1,420 MHz marker.[2] These targets were chosen because they represent relatively sun-like stars and, therefore, were seen as potentially likely to have Earth-like planets orbiting them. After approximately 150 hours of intermittent observation, and a false-positive signal caused by a secret military facility, Drake found no evidence of extraterrestrial technology (Michaud 2007: 35). The failure to produce any evidence of ETI did not deter Drake and he went on to become one of the most prominent figures in the scientific search for extraterrestrial intelligence.

A year later, Drake held the first scientific conference dedicated to SETI, gathering together a group of scientists in the physical, biological, and social sciences. This group included J. Peter Pearman of the National Academy of Sciences

[2] A common assumption of SETI has been that a signal might be broadcast at 1,420 MHz, which is the natural marker frequency for hydrogen, which, in turn, is the most abundant element in the universe. A reasonable assumption is that an alien intelligence sending out a signal with the intention of it being intercepted by another civilization might pick this frequency as it would represent evidence of intelligence—awareness of the importance of hydrogen in the universe.

space science board, Melvin Calvin, whose Nobel Prize was announced during the conference (Shuch 2011: 16) and NASA's Shu Shu Huang, who had developed the idea of habitable zones around stars in 1959 (Grinspoon 2009: 294), as well as Morris, Coccini, and a young Carl Sagan. The agenda for the meeting included a wide range of topics from stellar and planetary formation, the emergency of life, and the existence of planets within habitable zones. It also included themes surrounding communications technology, the evolution of intelligence, and the likely longevity of technological civilizations (Shuch 2011: 16). What became clear from the meeting was that the field of SETI research would have to encompass a very wide range of areas of intellectual investigations that included not only the natural sciences but also the social sciences and engineering.

One of the questions the participants realized represented a problem in thinking about how to pursue SETI research was the lack of knowledge we have about the abundance, or lack thereof, of extraterrestrial civilizations in the galaxy. To address this, Drake took the seven agenda items that had guided the conference and strung them together in an equation intended to estimate N, the number of communicative civilizations extant in our galaxy. The equation reads:

$$N = R f_p n_e f_\ell f_i f_c L$$

where

N = the number of civilizations in our galaxy with which we might be able to communicate
R = the average rate of star formation per year in our galaxy
f_p = the fraction of those stars with planets
n_e = the average number of planets that can potentially support life per star that has planets
f_l = the fraction of the above that actually go on to develop life at some point
f_i = the fraction of the above that actually go on to develop intelligent life
f_c = the fraction of civilizations that develop a technology that releases detectable signs of their existence into space
L = the length of time for which such civilizations release detectable signals into space

The Drake Equation has undergone some modifications in defining the variables over the years (Michaud 2007: 55), but it continues to provide a framework for thinking about what things we need to know in order to estimate the number of worlds in our galaxy that might harbor intelligent life. Some of the variables in the equation are fairly well understood. For example, over the past decade a variety of new tools, such as Galaxy Evolution Explorer (GALEX), the Spitzer Space Telescope, and the Hubble Space Telescope have allowed for new and more sophisticated observations related to the nature and rate of star formation in our galaxy. This has led to a recognition that stars do not form at a consistent rate: low mass and large mass stars form at differential rates related to the stellar environment (type of gas cloud) in which they form. And an important discovery of the GALEX mission was the detection of low-level star formation in environments that had previously been thought to lack star formation (Kennicutt and Evans 2012: 576). Kennicutt and Evans (2012: 595) make the important point that as "major observation advances are made, observational pictures that once seemed simple and certain have proven to be more complex and uncertain." Indeed, there have been somewhat competing models of star formation in the Milky Way and different opinions about whether the rate is continuous or occurs in bursts separated by long periods of relative inactivity (de la Fuente Marcos and de la Fuente Marcos 2004: 498). Nonetheless, current estimates are that our galaxy produces about seven new stars per year and about two solar masses per year ± 1.[3]

While we have a good understanding of the rate of star formation in our galaxy, it is also clear that the rate has not been consistent throughout the history of the galaxy; the number must have been higher in the past since our galaxy has converted most of its gas into stars. One question we don't know the answer to is how the rate of star formation might influence the rate of the formation of planets that could harbor life, but this information is important for arriving at a good value for R because it is clear that R is variable in relation to time and the numbers that might fit into this part of the equation can range from 6 to 24, which would significantly influence what we get for N.

New observational instruments such as the Kepler Space Telescope have allowed astronomers to develop a much better understanding of both f_p and n_e, the proportion of stars likely to have planets orbiting them and the average number of planets that can potentially support life as we understand it. Astronomers using data collected by Kepler have calculated a probability of about (34 ± 14 %) of stars like our sun hosting planets similar to Earth that orbit within the habitable zone although further study of the data is likely to generate greater precision in this estimate in the future (Traub 2012). The exact numbers are not really important to our discussion here. With at least 100 billion sun-like stars and possibly many more in our galaxy and recent estimates based upon Kepler data that about 17 % of these host earth-sized planets in close orbit,[4] it is clear that our galaxy has an abundance of potential hosts for extraterrestrial civilizations that might be in some way like our own. Indeed, some estimates indicate that there may be as many (or even more than) 17 billion planets in our galaxy that are in some way similar to Earth. Of course, this does not mean that these planets have Earthlike properties that could support life as we understand it—but it does suggest that there are a lot of possibilities out there.

The rest of the variables in the equation require considerable speculation on our part. For example, the fraction of planets that actually go on to develop life at some point (f_l,) is a variable for which we have only one data point—Earth. We have absolutely no idea how many other planets go on to develop life and we have no conclusive evidence that life has emerged anywhere other than Earth. A common response to this is that given the staggeringly vast size of our galaxy, let alone a universe with billions of galaxies, it seems unimaginable that our Earth is the only place on which life developed. I tend to agree with this position, but it actually is not logically tenable without empirical evidence—rather it is simply speculation. At present, we know of only one planet on which life has emerged, and we live on it. If the

[3] http://www.nasa.gov/centers/goddard/news/topstory/2006/milkyway_seven.html

[4] http://www.space.com/19157-billions-earth-size-alien-planets-aas221.html

probability of life emerging is so infinitesimally low that it only happened on one planet, then we just happen to live on the one planet where it happened. I don't think this is likely, but from a logical perspective, the possibility cannot be discounted—it is a potential reality.

The same can be said for the fraction of planets with life that go on to develop intelligent life (f_i) and those that might develop a technology that releases detectable signs of intelligence into space (f_c). Since we don't know the probability of life forming in our galaxy, we have no idea how likely it is for life to develop into intelligence, nor how likely it is for that intelligence to develop *the kinds of technologies we might be able to detect*. In other words, even if there are technological civilizations out there, do they use the same forms of technologies and types of communication that we do? Maybe they are intelligent bats with different types of sensory apparatuses and, thus, have developed a different technology for communication that we are unable to detect. I will return to this issue later. And, of course, the last variable in the equation (L), which represents the length of time that the civilizations from the variable f_c release detectable signals into space, could be rendered moot if the aliens don't use a technology we can detect. For now, I want to give some thought to the question of how likely it is that there are other civilizations in our galaxy.

In a book entitled *Rare Earth: Why Complex Life is Uncommon in the Universe* (2000), Peter Ward, a paleontologist with interests in astrobiology at the University of Washington, and Donald Brownlee, a University of Washington astronomer, argue that emergence of complex multicellular life on Earth, and the subsequent emergence of intelligent life, required a highly improbable combination of both geological and astrophysical events and conditions. Thus, they hypothesize, complex extraterrestrial life that requires an Earth-like planet should be extremely rare. In their Preface to the paperback edition, the authors make a point that is worth quoting at some length, stating that the widespread popularity of cultural phenomena like Star Wars, the X-Files, and Star Trek point to the fact that our society is enamored of and biased toward the idea that there is not only life, but intelligent life, throughout the universe.

> This bias toward the existence elsewhere of intelligent life stems partly from wishing (or perhaps fearing) it to be so and partly from a now-famous publication by astronomers Frank Drake and Carl Sagan, who devised an estimate (called the Drake Equation) of the number of advanced civilizations that might be present in our galaxy. This formula was based on educated guesses about the number of planets in the galaxy, the percentage of those that might harbor life, and the percentage of planets on which life not only could exist but could have advanced to exhibit culture. Using the best available estimates at the time, Drake and Sagan arrived at a startling conclusion: Intelligent life should be common and widespread throughout the galaxy. In fact, Carl Sagan estimated in 1974 that a million civilizations may exist in our Milky Way galaxy alone. Given that our galaxy is but one of hundreds of billions of galaxies in the Universe, the number of intelligent alien species would then be enormous (Ward and Brownlee 2003, xvii–xviii).

The point here, of course, is clear: maybe all of this talk about intelligent civilizations and the Drake Equation is just wishful thinking generated out of a cultural milieu in which there is a kind of feedback loop between science fiction and the thinking of astronomers deeply interested, and emotionally invested, in finding extraterrestrial life. Perhaps the conditions needed for the emergence of complex—note that Ward and Brownlee are talking about complex life, not even intelligence—are so unusual that there are only a very few civilizations peppered throughout our galaxy. If that were true, it is very unlikely, given the vast distances involved, that we would ever run into each other.

Where the Drake Equation gets hung up is on the fact that a large proportion of the variables require extreme speculation. Proponents of the equation argue that as our technology improves and we have increasingly sensitive methods for observing other stars and planets, we will gradually be able to input good data for the variables and should eventually be able to come up with a reasonable estimate of how many communicative civilizations are out there for us to greet. This will work even if the number turns out to be very low. My perspective on this is that Drake Equation is an interesting way of thinking about some of the variables we need to know about before we can make an educated guess at the number of communicative civilizations with technologies like ours that might be out there, but right now, throwing in numbers that are "educated guesses" doesn't do us any good, because the guesses aren't really educated since we don't have the data. It may be a long time before the equation becomes genuinely helpful in thinking about the number of civilizations out there.

That said, Ward and Brownlee raise a very important point about the influence of the cultural milieu on the formation of the Drake Equation and the ways its potential importance have been interpreted by astronomers and others. In many ways, the Drake Equation is as much a product of our culture as it is of scientific inquiry. It works from a set of assumptions about the abundance of intelligent life and nature of intelligence and of "civilization" that are, in fact, problematic. Ward and Brownlee have done a good job of raising some of the issues related to the abundance of intelligent life in their work; here and in the next chapter I want to focus on the manner in which both intelligence and civilization are constructed within the SETI community. One of the most significant problems with the Drake Equation is that it doesn't address the issue of what a civilization is—or even what a communicative civilization is. The assumption is that we're an example, which is a bit reminiscent of the Social Darwinists and their assumption that Europe of the nineteenth century represented civilization.

3.3 Civilization?

What, exactly, is a civilization? This word is constantly used in the SETI community as a label for the alien other that we hope to encounter, but rarely is a great deal of thought given to the concept of civilization as it is used here on Earth or in relation to those things that might inhabit other planets. Note, I didn't write people. I didn't write individuals. I didn't even write beings. I wrote *things*. This is part of the problem: What, exactly, do we expect to find when we bump into ET? Would a "civilization" for a species of highly intelligent dogs look anything like what we call civilization?

I will return to this point in the next chapter; before getting there, however, I want to expend a bit of ink on the issue of civilization.

Actually, if you look at the bookshelves of any library, you will find an enormous number of volumes with the word "civilization" in the title. When I searched the University of Texas library 493 titles came up with the word civilization in them, and I only used the American English spelling. Often the word is used to refer to the history of a particular, vaguely defined, group of people who are seen to have a common cultural history—e. g. *Western Civilization*, that boring, and highly misleading, title of so many college history classes. What, exactly, is Western Civilization? What are the boundaries of "the West?" When did its civilization begin? Do the Greeks and Norwegians really have enough in common to say they are part of the same civilization?

In their letter to readers at the beginning of their book on Louis the XIV, Will Durant and Ariel Durant (1963) note their frequently used definition of civilization as "social order promoting cultural creation." I'm not entirely clear on what this means, but the Durants go on to clarify that whatever civilization is it includes government, religion, science, music and art, moral structures, education, etc. They treat civilization in a very holistic way, which turns out not to be terribly helpful analytically since civilization ends up being everything humans do from a social perspective. Furthermore, these definitions include every form of organized human group throughout our history. Neanderthals had social order promoting culture and also appear to have had something akin to religion (funerary ritual), art (they made jewelry), social organization through which they governed themselves, and had sufficient science to build stone points and other tools (Papagianni and Morse 2013). But the Neanderthals are not a group that we normally equate with civilization. A more useful definition of civilization can be found in Wikipedia (believe it or not):

> **Civilization** or **civilisation** generally refers to polities which combine three basic institutions: a ceremonial centre (a formal gathering place for social and cultural activities), a system of writing, and a city. The term is used to contrast with other types of communities including hunter-gatherers, nomadic pastoralists and tribal villages. Civilizations have more densely populated settlements, characterized by a ruling elite, and subordinate urban and rural populations, which, by the division of labour, engage in intensive agriculture, mining, small-scale manufacture and trade. Civilization concentrates power, extending man's control over both nature, and over other human beings.

This definition provides an emphasis on structural elements of social organization and power relationships and contrasts civilization with societies that may lack physical and organizational technologies associated with large-scale populations as well as the tendencies to employ power structures and build social hierarchies that are common among large-scale societies that are much more extensive and stratified than what is typically found in human societies such as bands or tribes.

The idea of trying to define civilization by those who thought they were living in one actually has a pretty long history and this tends to create rather biased ideas about how we should define it. Scholars in the nineteenth century often had a very clear sense of what civilization is and what it is not. American lawyer and early anthropologist Lewis Henry Morgan, like other social Darwinists of his time, developed a scheme for classifying human forms of social organization in such a way that

Table 3.1 Morgan's evolutionary model of human social organization from savagery to civilization

Type of society	Technological features
Lower Savagery	Fruit, nut subsistence
Middle Savagery	Fish subsistence and fire
Upper Savagery	Bow and arrow
Lower Barbarism	Pottery
Middle Barbarism	Domestication of animals (Old World), cultivation of maize, irrigation, adobe and stone architecture (New World)
Upper Barbarism	Iron tools
Civilization	Phonetic alphabet, writing

civilization is a product of cultural evolution over long periods of time in much the same way many at that time (as well as now) saw biological evolution to lead to more complex and advanced organisms. Morgan views human history as a process of evolution through three major "ethnical periods" he identifies as savagery, barbarism, and civilization. The lower of these—and yes Morgan and Social Darwinists like Spencer and Tylor see these as primitive, less progressed forms of social organization—can be divided into smaller stages so that we get this sort of schema for the evolution of human societies (adapted from Harris 1968: 181) (Table 3.1).

The important thing to recognize in this approach to thinking about human social organization is that it has a *telos*. By this, I mean that it's seen as directional and assumes the *improvement* of humans over time; that improvement being expressed in both better forms of social organization and associated better forms of technology. And for Social Darwinists like Morgan and Herbert Spencer the rise of civilization also brought with it improved *moral* capabilities in humans. Spencer (1873: 45–46) argues that humans, as a part of Nature, are constantly immersed in a process of evolution in which each "successive result becomes the parent of an additional influence, destined to some degree to modify all future results." Progress, for Spencer (1873: 80), is not an accident, but is a basic feature of Nature that is also evident in the ongoing changes in human technologies and social structures that result from "a law underlying the whole organic creation; and provided the human race continues, and the constitution of things remains the same, those modifications must end in completeness… so surely must human faculties be moulded [sic] into complete fitness for the social state; so surely must the things we call evil and immorality disappear; so surely must man become perfect."

Indeed, Spencer sees human social or cultural evolution in terms of progress that is directional and within which certain "primitive" kinds of behaviors and attitudes are historically prior, *uncivilized* and ultimately morally inferior. Uncivilized humans, for Spencer (1873: 449), must be "devoid of sympathy, or must have but the germ of it, for he would otherwise be incapacitated for his destructive office. In other words, he must be what we call a savage, and must be left to acquire fitness for social life as fast as the conquest of the earth renders social life possible. Whoever thinks that a thoroughly-civilized community could be formed out of men qualified to wage war with the preexisting occupants of the earth—that is, whoever thinks that men might behave sympathetically to their fellows, whilst behaving unsympathetically to inferior creatures, will discover his error on looking at the facts." Humans, in their baser uncivilized forms, are cruel; they do things like torture, cockfights, and the Roman gladiatorial games. In contrast to this, Spencer tells his readers that, "amongst ourselves [read civilized Europeans] a desire to diminish human misery is accompanied by a desire to ameliorate the condition of inferior creatures" (1873: 450). Yes, we all know that nineteenth century Europeans never tortured anyone—they were a really nice and devoted to making the world a better (read European) place.

Of course, the evolutionary schemas of Social Darwinists were discredited long ago. Anthropologists such as Franz Boas and his students wrote extensively and convincingly against the racist anthropology of thinkers like Spencer and Morgan. In his book *The Mind of Primitive Man*, Boas (1921) denies the idea that there are innate differences between "savages" and "civilized" humans. It was history and culture that led to differences in custom and practice, rather than biological differences that were expressed in higher and lower forms of social organization (Cartwright 2000: 22). Boas clearly recognized that although social change is obvious and as populations increase we see increasingly complex forms of social organization, normally with increased layering and specialization of social functions and occupations, there is nothing *better* about one form of human social organization over another. In short, change in human social organization does not imply improvement or progress.

This is important for our discussion here because in many ways it is the social Darwinist concept of civilization that lurks in the assumptions of SETI researchers when they use this term or when they talk about how human civilization will compare with that of an extraterrestrial intelligence. I want to be careful here, because I am not claiming that SETI scientists are closet Social Darwinists. Rather, the manner in which they tend to conceptualize civilization draws upon assumptions about social and cultural evolution that have the same sort of *telos* found in the ideas of the Social Darwinists.

For example, former director of the SETI Institute, Jill Tarter, has argued that our extraterrestrial interlocutors are likely to have *outgrown* organized religion. Tarter develops this idea from the oft-cited notion that any extraterrestrial civilization with which we might come into contact is necessarily going to be much older than our own. Why? Well, it has to do with both technological development and social stability over time. Tarter (2000: 145) states that, "any other technology that we detect in the near future will be more advanced, simply because we could not detect any technology more primitive than our own, and the probability that their technology would be exactly at our stage of development is vanishingly small." This brings us back to the Drake Equation and to the variable L, the length of time a civilization has been releasing detectable evidence of technology into the cosmos.

The argument goes that any extraterrestrial technology that might be detected by our "primitive technology necessarily will be far older than we are" (Tarter 2000: 145). By far older, Tarter argues that L is likely to be measured in the tens of millions of years for any solar-type star that is relatively nearby, meaning within 1,000 light years or so. This means that the technologies, and the civilizations that make those technologies, must be very long-lived and, therefore, must have developed a stable form of social structure that is unlike what we currently have on Earth. Beyond this, according to Tarter, said long-lived civilization is likely to either never have had, "or have outgrown" organized religion. Finally, Tarter (2000: 148) argues that as long as contact with us does not somehow threaten the "continued longevity/stability" of their society, ET is likely to "act in our best interests." In other words, ET is likely to be altruistic unless we appear to be a threat, which seems unlikely since we will be comparatively primitive.

There are quite a few problems with this argument. First, the types of terms Tarter uses to describe us and them are reminiscent of the Social Darwinist perspective. This is most notable in the idea that an extraterrestrial civilization would have likely "outgrown" organized religion. Our experiences here on Earth suggest that as civilization has become more complex and stressful, religion has actually *increased* in strength even while religions like Christianity have been challenged by scientific knowledge. And, then, we get back to the problem of defining religion. Would an advanced civilization that had an organized religion like Buddhism have any real need to outgrow it? Buddhism is quite amenable to the idea of modern science, so why would it be necessary to shed the spiritual side—organized religion and science are not inherently in conflict in the Buddhist world. In Tarter's perspective, we have a rather ethnocentric conceptualization of organized religion that significantly weakens her central premise.

In the end, there is no way to know if a far more advanced civilization would or would not have organized religion—perhaps they would actually be a theocracy—but the evidence on Earth does not support Tarter's hypothesis. The most serious problem with this proposition lurks in her use of the term "outgrown" which points to an assumption that as civilizations evolve they get better, they progress. This is a concept of social and cultural evolution that reflects the same type of teleological assumptions found in the Social Darwinist model, even if it lacks the kind of racism that was implicit in those models. Elsewhere, Tarter,[5] argues in response to Stephen Hawking's claim that there is a reasonable chance ET will be very nasty, that "a large value for L (a requirement for that magical, star-spanning technology) could also mean that their distant civilization had found a way to stabilize itself in order to survive and grow old. That might require outgrowing any aggressive and belligerent tendencies that may have characterized their youth." Again, we see similar types of assumptions in the notion that aggressive and belligerent behavior characterizes the "youth" of a civilization, while altruism and general niceness hopefully will characterize the qualities of a long-L advanced extraterrestrial civilization.

Tarter may be right, but we don't have any empirical evidence to support this claim. In fact, I would argue that virtually all of the evidence throughout the history of Earth is precisely the opposite. The larger and more technologically advanced in comparison to its neighbors a society gets, the more belligerent it tends to become. The U.S. is a pretty good example of this in the modern world and in the past 10 years or so we have seen this emerge very clearly with China. As China has become

[5] http://www.cnn.com/2010/OPINION/04/27/tarter.space.life.fears/index.html

more economically powerful, it has become increasingly intransigent and even belligerent with its immediate neighbors like Japan and Taiwan.

There are other assumptions lurking in this formulation of extraterrestrial civilization. In terms of longevity, Tarter assumes that the rate of social evolution, and thus technological innovation, is constant throughout the galaxy and that these two are necessarily tied together. This is the same basic assumption about the linkage between technological advancement and moral advancement that we see among the Social Darwinists of the nineteenth century. However, as noted in the previous paragraph, nothing on Earth would suggest that improved technology is accompanied by altruism, increasingly responsible behavior, or just being nice. Let me make this very clear: *there is no evidence on Earth that technological advancement is in any way linked with moral or social advancement.* At the core of this problem is that there is no agreed upon basis among humans for determining what is inherently morally good behavior or even simply advanced cultural practice. In the U.S., we tend to believe that egalitarianism is a moral good; our ideology emphasizes the notion of equality for all and frames that idea in terms of our society being *better* than societies that don't value equality or at least don't value it in the same way we do. This is a basis for viewing the US as a politically (and morally) advanced society. However, by this measure, tribal and band societies are morally superior to state level societies like ours because decision-making is usually diffuse and shared and there is very limited differentiation in wealth and social status. The point here isn't really which type of social organization is better or more advanced. In fact, it's relative; there is no one form of social organization that is clearly better than other forms. Rather, the point is that there are differences and there is no necessary linkage between technological advancement, complex social structure, and morally good behavior. And when it comes to behaviors and concepts like altruism, we are really talking not about simple biological elements of behavior,[6] but about conceptual categories about moral behaviors that are highly shaped by culture. One society's altruism may not be the same as that of another society and altruism itself has the potential to shift from being a positive to a negative influence on human behavior, as has been well shown in the work of Barbara Oakley and others doing research on pathological altruism (Oakley 2013; Oakley et al. 2012).[7]

Tarter's ideas are by no means unique in the SETI community. Many astronomers and others have discussed, debated, and argued in favor of the notion that ETI

[6] I recognize that there is a biological basis for altruistic behavior that has been widely discussed in the areas of sociobiology and physical anthropology. I am not disputing this, rather I am arguing that the biological aspect of altruism is heavily shaped by culture. See Traphagan (2012).

[7] There are those who argue that historical characters like Hitler were actually practicing a form of altruism. For example, in an opinion piece published in MIT's The Tech, David Honig argues that, "I claim that Hitler was altruistic… He required individuals to put the good of other people, whose will was represented by the state, above their own." I personally find it difficult to see Hitler as altruistic. My point is that the concept of altruism is difficult to define in a universal way and it is possible to conceptualize actions that many view as monstrous as being altruistic depending upon your viewpoint. This isn't simple. http://tech.mit.edu/V105/N53/honig.53o.html

will be morally superior to humans and that this is linked to the time scales and technological progress associated with long-L societies. Frank Drake has strongly supported the notion that we should expect altruism form ET:

> Should altruism be expected, perhaps be ubiquitous, even universal? There seems to be an easy answer to this question—yes. As noted by many writers, altruism can be expected in intelligent creatures simply as a Darwinian imperative. Communities of mutually supportive individuals, practicing altruism, will possess greater potential for survival than the same individuals acting alone. Even greater survivability accrues when individuals have the will to endanger their own well being for the good of the community.[8]

A deep analysis of culture and history on Earth shows that the answer to the question Drake poses at the beginning of this paragraph is anything but an easy "yes". How one defines the good of the community is cultural. And even from a biological and evolutionary perspective altruism is anything but a simple feature of human behavior. Barbara Oakley (2013: 10409) has made this very clear in her work on pathological altruism. While recognizing the evolutionary advantages and social benefits that come with a biological bias for altruism, she also points out that altruism involves tradeoffs that can take a decidedly negative turn.

> [P]athological altruism can be thought of as a pattern of nurturing or beneficial behavior with evolutionarily unsuccessful consequences. Evidence for antecedents of such behavior can be seen in the animal world; examples include the unwitting hosts of brood-parasitism, as with the wood thrush who devotes substantial resources to raising the offspring of cowbirds. Such antecedent behavior is manifest at even a genetic and molecular level. For example, beneficial replication processes within a cell can be co-opted by viruses. Consequent cell lysis or exocytosis allows the new viral bodies to spread the contagion.

In short, altruism is not an unproblematic good that we should necessarily expect to find in morally and socially advanced societies. And the fact is that humans have been doing "civilization" for thousands of years and we have not outgrown our tendency to be belligerent and nasty, despite the fact that we probably would have better lives if we cooperate and act altruistically towards each other.

There is a final assumption that I want to raise that floats around in the writings of astronomers like Tarter and Drake. That is the idea that stability and social unity are likely to be present in advanced civilizations and that they are necessary for the survival of a civilization (whatever that means) over a long period of time. The assumption here is that an alien civilization is not only stable, but unified and that there is only *one* stable civilization on a given planet that might be sending out signals we could detect.

Let's turn this around for a minute. Imagine that you are an extraterrestrial alien living on another planet looking up into the cosmos wondering if there is anyone else out there. You decide, with a bunch of your similarly fascinated colleagues to build a device that might detect the presence of alien technology in the form of radio transmissions. Maybe you call the device the Glerzmurf Radio Telescope Array (GRTA) or something like that. You start listening and after a few years you get

[8] Encoding Altruism, from the newsletter *Science and Spirit*. http://archive.seti.org/seti/projects/imc/encoding/altruism.php

really lucky and you pick up a transmission that is clearly not from your planet and shows signs of intelligence—well, that's relative, because it's actually a broadcast of the Flintstones from 1965 on Earth. Of course, you don't know this, but you do figure out where the signal came from and in fact your scientists have sufficient technological prowess to determine that it is from a small, rocky planet that orbits a medium sized star about 50 light years from your own world.

This is really cool. Everyone gets excited. Representatives of the news media come to interview you. You become famous, hooking into the local scientific paradigm that discovery of something novel and new is the goal of science, because you found evidence of another intelligent civilization (with really weird cars) and one that isn't even very far away by cosmic standards! Scientists and politicians—let's assume they don't have religion, so we can leave out theologians—gather and decide what to do. Do you send a message to the civilization that lives on that other world? And what do you say? Conferences convene, debates ensue, op-ed pieces are written, all about whether or not to send and what to say to the aliens on that other little world.

Then an annoying anthropologist stands up and asks a question: Does that world have *a* civilization? Or does it have *many* civilizations? And if it has many, how do you frame a message? What if they aren't unified like us? We gave up organized religion eons ago, but there was a time on our planet when some religious sects were terrified of the idea that there might be extraterrestrial civilizations, because they thought it would challenge their belief that we were at the center of the universe. It's great that we matured past that, but how do we know that it is necessarily a requirement that a planet mature past that in order to develop a civilization or maybe multiple civilizations?

Which is it? Does Earth have one civilization or many? Will all parts of our "civilization" react in the same way? The answer to these questions depends upon how we think of civilization and also what we think the influence of contact might be on relations among groups of people. Is civilization a planetary type of social structure, or a kind of social organization that occurs in many different ways at a given time and throughout history? The answer will also deeply influence how we conceptualize the alien other from whom we have received a transmission. Well, they sent this transmission, so they must be pretty advanced (although the transmission was pretty weak). Maybe they sent it intentionally—maybe they detected a transmission from our planet and they are trying to reply. If that's so, then they must be highly civilized—a unified, stable society, that has achieved a peaceful existence. Maybe they are pretty nice, too. Hmm.

I don't write this to be facetious. This is a reasonable course of thinking, because if we think about our own world—*and that's the only basis for discussion we have*—then we need to ask these questions. I realize that SETI astronomers will remind me that the likelihood of obtaining a transmission from a short-L source like our own planet is very low. The problem here is not with the age of the society but with the idea that the longer a civilization exists the more likely it is to be unified and stable. This is *not* evident in our own history and there is no necessary reason to think that it would be true of a long-L civilization from another star system, either. In fact, the history of our own world is one that includes the regular *rise and fall* of civilizations. On what grounds should we assume that this is going to stop anytime in the foreseeable future? Is it, in fact, likely that a highly advanced alien civilization will be either unified or stable over long periods? Perhaps that world goes through periods in which there are high levels of technological innovation punctuated by long periods where not much happens and their society or societies destabilize for a period of time—a social or cultural form of punctuated equilibrium. In fact, if they had managed to get off of their planet and colonize even one other planet in their star system, then they would have done a pretty good job of reducing the risk of complete collapse of their society, even if conflict and major fluctuation continue.

The typical SETI approach to thinking about civilization is linear in the same sense that it was among Social Darwinists. Civilization follows a straight path of progress and improvement leading to a superior society characterized by unity, stability, moral superiority, and lack of belligerence. But the empirical data we have on Earth don't confirm this. On Earth, civilizations rise and fall, technologies come and go—think about the Romans who invented concrete and used it widely from around 300 B.C. to 476 A.D. only to have the technology forgotten for close to 1,000 years. Here is another way to think about this. Suppose that 50 years from now, after 150 years of regular transmission of radio signals into space, Earth experiences a massive plague or an environmental disaster and our "civilization" or "civilizations" collapse. We stop transmitting for several decades or centuries. But over time we rebuild and are again transmitting. This is pretty much the story of Walter Miller's post-apocalyptic novel *A Canticle for Lebowitz*. Does that seem unlikely? I don't think so. Does it represent stability? I don't think so.[9]

The point I want to emphasize in closing this chapter is that the notion of stability and moral superiority of extraterrestrial civilizations belongs much more to the realm of science fiction and conjecture than it does to science fact. It is highly speculative and not grounded even in the empirical evidence we have from our own planet. And the very notion of civilization itself may turn out to be quite different among creatures that neither look, nor think, like us. To that issue we now turn.

[9] There is another point that needs to be kept in mind. SETI scholars often describe our civilization as in an "adolescent" stage, as I have noted. But depending upon how you look at it, we have had civilization on Earth for around 11,000 years, dating back to the Late Natufians who developed agriculture somewhere between 13,000 and 11,000 years ago (see Tomkins 1998).

Chapter 4
Culture, Intelligence, and ETI

Even if we never reach the stars by our own efforts, in the millions of years that lie ahead it is almost certain that the stars will come to us. Isolationism is neither a practical policy on the national or cosmic scale. And when the first contact with the outer universe is made, one would like to think that Mankind played an active and not merely a passive role—that we were the discoverers, not the discovered.

— Arthur C. Clarke, The Exploration of Space, 1951

Well, maybe. But the fact is that the comparison of earthly and cosmic scales is a bit absurd. Given our current understanding of physics and our lack of knowledge about the probability that there is anyone out there with whom to make contact, isolationism may be our only option. However, there is value in speculating about both the nature of ETI and how we might make contact, for engagement in such conjecture tells us much about ourselves as humans and the ways in which we think about self and other. The project and process of thinking about ETI raises several important questions that not only have significance for our understanding of ETI, but also are related to the development of morally ambiguous technologies such as AI and genetic engineering. When we contemplate the nature of and communication with ETI, we necessarily must ask: What is intelligence? What is culture and how does it influence the expression of intelligence? How do our cultures and biology influence our capacity to communicate and our ability to recognize intelligence in non-human beings (or in other human beings for that matter)?

Recent research focusing on how humans might construct interstellar messages to communicate with an extraterrestrial intelligence has raised interesting opportunities to think about the manner in which contact and culture intersect. Vakoch (2000), for example, asks the intriguing question of whether or not music—the seemingly universal "language" shared by humans—could provide a means of communicating

J. Traphagan, *Extraterrestrial Intelligence and Human Imagination*,
Space and Society, DOI 10.1007/978-3-319-10551-2_4

with ETI or whether it might be incomprehensible to beings with different types of sense organs and who evolved potentially incommensurable ways of dealing with the exchange of information through sound. It has also been widely thought that mathematics would provide an excellent means of communication (Letaw 2011: 176)—in order for extraterrestrial intelligent beings to construct the equipment necessary for radio transmissions, it is reasonable to assume that they would need a thorough understanding of mathematics. And other researchers have suggested exploration of potential cognitive universals such as aesthetics or spiritual ideas as a basis for developing strategies in SETI research and the construction or interpretation of interstellar messages (Lemarchand and Lomberg 2011: 371). However, in these cases, questions arise about whether or not the symbolic systems associated with mathematics, aesthetics, or spirituality of an extraterrestrial being will be mutually intelligible with our own, even if the underlying principles are the same. Perhaps most interesting are the thoughts of mathematician Carl DeVito on the universality of mathematics. While not dismissing the idea that mathematics might work, he points out the considerable difficulties in using even mathematics as a universal language, because rather than representing a "real" world, mathematics represents an "artificial world, a world of abstractions and idealizations that human mathematicians have created over many centuries." Mathematics, according to DeVito, may be much more a reflection of our minds than it is of an objective reality independent of those minds (DeVito 2011: 440 Vakoch 2001). DeVito's very important point is that we must be careful not to assume that things we believe to be representations of an objective reality actually are themselves objective or universally intelligible representations of that reality.

Another way to think about ETI is to ask what the basic components of intelligent beings ought to be, given what we know about life on Earth. Cognitive scientist and AI pioneer Marvin Minsky (1985) argues that certain basic capacities and characteristics will be typical of any intelligent being regardless of where they happen to have evolved: these include problem solving, analytical skills, the ability to describe the world, explanatory skills, accumulation and exchange of information, allocation of scarce resources, planning ahead, and self-awareness. While this makes a great deal of sense, there is an important issue missing that needs to be considered—all of these capacities involve culture and culture is highly variable, even among humans who are from a biological perspective a fairly consistent species (we don't show the diversity of sizes and shapes that dogs do). For example, on the one hand problem solving has a certain uniformity in that it requires some sort of rational process in order for the problem to be solved, but the method of problem solving can vary considerably from one society to another. Some people may place an emphasis on reaching consensus while others may emphasize majority rule or authoritarianism. Most problems that involve behavior and social organization have many solutions and the type of solution arrived at is in part a product of the approach to solving problems viewed as being normal in a given cultural context. It is not unusual for the approach, and thus the solution, of one group to be viewed as odd, irrational, or simply wrong by another group.[1] This type of complication would likely be intensified if we think about

[1] Anthropologist E. E. Evans-Pritchard's (1976) work among the Azande provides a very good example of this.

differences in approaches to problem solving between humans and an alien other whose sensory apparatus is likely to be quite different from ours, and thus its understanding of the world is likely to be quite different.

Philosopher Thomas Nagel's discussion of the question, "What is it like to be a bat?" is illustrative when considering this issue. Nagel (1974: 438) notes that while consciousness of experience occurs among various types of animal life, understanding what it is to experience *being* another form of animal life is essentially impossible. Bats, of course, have evolved to experience the world through echolocation using the sonic reflections of their own screeches to discriminate among objects in the world and to construct some type of model or understanding of their environment in their brains. Echolocation mediates the manner by which bats enact the capacities that they have, such as planning ahead to avoid ramming things or identifying mosquitoes to eat. In other words, *how* they plan and *how* they identify things is based upon a model of the environment in their brains that is itself based upon sensory organs that process and a brain that interprets sound reflections as the means by which to range, spatially locate, and identify objects. Of course, this is very unlike humans who primarily interpret light reflections to accomplish the same goals.

The difficulty raised when we think about bats is that because we lack the capacity to echolocate, we are fundamentally incapable of knowing how bats experience the physical world we both inhabit. We can *imagine* what it is to be a bat, but we cannot *know* what it is to be a bat or what a bat's experience of the physical world is really like, because we are incapable of processing and interpreting information in the way that bats do. Nor can we know if our imagination about bat experience actually reflects the ontology of bat experience. In short, while we inhabit the same physical world as bats, we do not inhabit the same mental world, and the mental world is the one that really matters when it comes to communication, because all communicative events are mediated through processes of cognition.

The same can be said for other animals, such as dogs, that are much closer than bats to humans in terms of their sense organs. Hound dogs have approximately ten times the scent receptors that humans do and have different visual and aural abilities. Dog noses divide the air into two streams, one that goes to their lungs and the other that is used for sensing the world, which gives them a very clear "view" of the world that they smell. With these senses, how does a dog construct the world? Are his "images" or conceptualizations of the world primarily based upon interpretation of scent or sound as opposed to the interpretation of visual images? Interestingly, it is so difficult for us to imagine the dog's world, due to our being largely visual animals, that we are forced to resort to words that index seeing (image, imagine) when we describe how a dog or any other being might think about the world. My guess is that the "images" dogs create in their heads are quite different from those we create in our heads. And we do know that even when it comes to images of the visual type, dogs don't see the world as we do. They can only see the colors blue, violet, yellow and some shades of gray. If you give a dog an orange toy, she may not have much interest if that toy is set among a bunch of other toys that are, say, blue, because she can't see the orange toy as clearly. Her perspective on what to focus on visually in the world is limited by her physical ability to see color, just as ours is—we don't see things in infrared.

Even among humans, who largely share the same apparatus for sensing the world (keep in mind that people who are blind or deaf, for example, do not share the same apparatus), the things that people choose to focus on when constructing their world can vary considerably from one culture to another. For example, when English speakers count, we focus only on the number of a particular object—one sheet of paper, two sheets of paper, three sheets of paper, one pencil, two pencils, three pencils, etc.—and emphasize the difference between one object and multiple objects. For some strange reason it matters to English speakers that when counting things one only addresses the issue of "how many" and that when one categorizes parts of the world one differentiates between a single object of a particular kind and many of those objects.

By contrast, Japanese speakers approach counting things quite differently. First, the Japanese language does not need to distinguish between one and more than one object—this is obvious by the fact that numbering of things does not involve plurals in most cases. There is one car, two car, three car, or one tree, two tree, three tree, and so on. Second, the issue of how many is not distinct from the issue of the structural form that the object one is counting takes in the world. Thus, to count things like sheets of paper or compact disks in Japanese, one counts 一枚、二枚、三枚. But perhaps you do not read Japanese. In Romanization, these are spelled *ichi-mai, ni-mai, san-mai*. If you look at the kanji, you will see that the first character in each is actually quite easy to figure out—one stroke for 1, two for 2, and three for 3. Unfortunately, that's where the fun ends, because the numbers 4, 5, 6, 7, 8, 9, 10 look like this: 四、五、六、七、八、九、十. It is the other character, 枚, that turns out to be more interesting for our purposes. This character, read as *mai*, consists of two component parts that mean tree (木) and hit or strike (攵), although this particular kanji character is not generally used as a stand-alone character, instead being primarily used as part of other characters.

So what does *mai* mean? It is a counter placed after the number and that refers to flat, thin objects. Thus, the way Japanese people count is to say, "one thin flat thing, two thin flat thing, three thin flat thing." And if we want to get very technical and precise, even the word "thing" is problematic, because it's assumed in *mai*. In fact, when you look up this character in a Japanese-English dictionary, the translation is usually something like "counter for flat, thin objects or sheets." Note that there is no single word in English that represents a translation of the word *mai* from Japanese. While we can capture the basic meaning of the word, a direct translation from the Japanese into the English is actually not possible. We simply do not have a way of counting in English that parallels the Japanese approach. In other words, there is a basic incommensurability in the paradigms of counting in Japanese as opposed to English, by which I mean that the terms for counting cannot be directly translated from one language or counting paradigm to the other and this leads to a certain amount of information loss in the process of translation.[2]

[2] There is a very large literature in philosophy that deals with the issue of commensurability in relation to scientific paradigms as well as in relation to language, but this discourse is too much of a diversion from the topic to explore here. I encourage you to look at works by Kuhn (1962a), and Feyerabend (2010) for a deep exploration of this topic, as well as a useful discussion of Wittgestein's ideas about commensurability by Glock (2008).

If one wants to count things like pencils or pipes in Japanese, one counts *ippon, nihon, sanbon,* indicating "one cylindrical thing, two cylindrical thing, three cylindrical thing." There are counters for large machines and vehicles (台), people (名、人—there is more than one for people used variously in relation to the word for people being used), large animals (頭), small animals (匹), years of age (才), taking a turn as in sports (番), tatami mats (畳), and so on. I think you get the point—there are a lot of these, and, in fact, there are more than 100 different counters.

The issue here is that even between two human languages, the approach to something as simple as counting differs significantly, although by no means beyond the point of mutual comprehension. We can translate counting in Japanese into English. When translators do this, they render "*enpitsu ippon*" as "one pencil" and "*kuruma ni-dai*" as "two cars." While this is a perfectly clear and reasonable translation for these objects and how many of them there are, something interesting happens in the process—basic interpretive and classificatory information associated with how Japanese people perceive what is important in counting things is lost. Something that is apparently important about how things are classified when counting does not transfer when we translate counted objects from English to Japanese and vice versa. Furthermore, we encounter this difference despite the fact that Japanese and English speakers do all of the things that Minsky identifies as being fundamental to intelligence.

How would we translate counting between humans and, say, intelligent beings who process sensory data through echolocation? Would counting—and more generally mathematics—necessarily be constructed in the same way humans do this by such beings? Perhaps such beings would be quite interested in shape and size or reflective qualities when counting, given their manner of processing the world—would a bat-like intelligent species count "one sound-absorbent thing, two sound-absorbent thing, or one sound-reflective thing, two sound-reflective thing?" Or perhaps the notion of discrete objects in the world would be less meaningful for echolocating creatures; maybe echolocating beings do not perceive the word as a collection of discrete objects in the way that visual beings like humans do. I'm not sure what it would be like to experience the world through echolocation, but I can say this: when I listen to things in the world, the notion of discreteness that I experience when I look at things is different. I can hear the differences in the notes and tones of instruments in a tune played by a jazz quartet, but my general experience tends to be quite unified—I have to concentrate to pull things apart and the musical experience is one that involves a blend and unification of sounds, not a separation of them. To a bat, are the objects that it hears discrete in a way commensurate with the objects that we see as being discrete?

Even while the underlying principles of symbolic systems such as mathematics or (maybe) music should be understood by both humans and an alien intelligence, the manner in which a particular being obtains and processes sensory data will influence the way in which it constructs any system to describe what is being processed. Furthermore, the elements of the world that are deemed important in a particular culture, which in turn are shaped by the sensory organs available to a particular species of beings, also will influence cognition and the manner in which individuals in that society classify and construct their world around elements that matter more or less.

Whether it is counting, music or mathematics, the question of how to communicate is not simply one of the medium through which ideas are exchanged or the capacity to carry out certain functions necessary to understand and manipulate an environment that is consistent in a universe that both beings inhabit. Instead, it is one of recognizing an interpretive context or framework for communication that will work when it is quite possible that the mode of communication and interpretation—of receiving and interpreting information—of an extraterrestrial life form will be significantly different from our own. At the root of this issue is the question of how culture factors into that process of communication across different intelligent species and how culture and biology intersect to construct reality.

4.1 What Is Culture?

I have used the term culture several times throughout this book without offering a precise definition. Part of my reason for this is simply that although culture is difficult to define, we tend to operate as though we know what it is—I wonder how many readers have at some point in the book asked themselves, "what does this guy mean by culture?" Hopefully, quite a few of you have, but my guess is that most have read the word without giving its definition much thought. We use culture regularly in casual conversation without thinking a great deal about what the "thing" *culture* actually is. Obviously, this is problematic when we want to think about and observe the influence of culture on issues such as interstellar communication or the nature of extraterrestrial beings from an analytical perspective—keeping in mind Feynman's concern about not overdoing it, we do need to be able to work towards definitions of concepts like culture if we want precision in analysis of what we encounter or sophistication in speculation about what we may encounter.

In general, the definition of 'culture' is assumed rather than explicated, not only in the literature dealing with SETI, but more generally in scholarly and non-scholarly discussions of human social organization and particularly in the area of cross-cultural communication. However, the culture concept is actually quite problematic even if we only think about it as a thing that humans or human societies have. Where is it? How do we identify it? And how is it a thing that all humans seem to have but also seems to be different everywhere we look?

Anthropologists have long recognized the difficulties associated with identifying the characteristics of any particular "culture" and have debated the extent to which one can consider culture a bounded "thing" that can be observed and analyzed, as well as the extent to which it determines behavior. In other words, culture is problematic both in terms of its use as an empirical category and as an analytical category. Debate about how we should use the culture concept has often divided anthropologists about what it actually represents in terms of human social organization and behavior. As Watson (1997) points out, early usage of the term in anthropology centered upon the idea that culture is a shared set of beliefs, customs, and ideas that are learned and unify people into coherent and identifiable groups. In this sense, then, culture represents a form of collective or social memory that links past, present, and future. This formulation takes culture as a relatively deterministic feature of human social organization that shapes behavior within a particular society, the edges of which seem to be clearly identifiable. Historically, these bounded entities often have been associated with the geographical space in which a particular group of people lived. This is what I call the bounded-culture model, in which culture is thought of as a fairly static thing that can be observed and recorded. It is this model that continues to be used widely in public discourse as well as in some areas of scholarly work such as cross-cultural communication when talking and writing about culture.

Contemporary anthropologists have come to view culture as much more fluid and vague than what we see in the bounded-culture model. The more one peers into the workings of culture, the more it becomes clear that people not only are held together, but may be divided by their customs and beliefs, even when they ostensibly belong to the same culture. Rather than a deterministic "thing" that shapes human ideas and behavior, this view of culture, which I call the process-culture model, represents culture as a changing web of ideas and symbols in relation to which people not only follow, but continually contest and reinvent the customs, beliefs, and ideas that they use, collectively, individually, and often strategically, to characterize both their own groups and the groups to which others belong (Ortner 2006; Prashad 2006). In short, culture is in a constant state of flux, which means there is no way you can grab it, hold it, and analyze it in a constant form. As soon as you look at some aspect of a particular culture, it has changed.

This should not be taken as meaning that culture is somehow not observable simply because it is expressed in the constantly changing abstractions of ideas and symbols, rather than in concrete things. It is observable in the sense that it can be *evaluated* using the techniques associated with ethnography that are typically employed by anthropologists. However, the properties of a particular culture are best understood as *latent* rather than *possessed*, by which I mean they are attributed conditionally by those engaged in observation of the culture or group of people who display patterns of behavior that we associate with a particular context of interaction.

To paraphrase the work of physicist/philosopher Henry Margenau (1949: 296), if we ask about the properties or features of a particular culture, our experience leaves the number indefinite; thinking deeply about the nature of a culture will show us that no limit can be ascribed to the features or characteristics of American culture, Japanese culture, French culture, etc. The reason for this is, in part, because each of these "things" are constantly in a state of flux—they are always changing as individuals in those contexts interact and create new ideas, symbols, beliefs, and behaviors. However, in conventional discourse we treat cultures as though a finite conglomeration of properties suffice to identify Culture X as Culture X. We use a crucial set of properties—often stereotypical properties—selected from a likely infinite number of potential properties to characterize a given culture at a given point in time. This occurs both in public discourse and in many areas of academic discourse related to identifying whichever culture we happen to be observing.

In our ordinary experience of the world, assuming that the properties of a given culture are possessed by that culture or, more accurately, by the individuals associated with the culture, is a convenient way of thinking about collectivities. It allows us to characterize groups of people in conventional ways that operate as a kind of shorthand to represent frequently observed behaviors and ideas of a defined group, the boundaries of which can be quite arbitrarily identified. However, an accurate understanding of culture works from the perspective that the attributes of a particular culture are generated through an interaction between observer and observed, rather than being inherent in the things observed—that is other people. A culture—meaning the ideas, symbolic structures, and concepts inside an individual's head that motivate behavior—does not have specific characteristics at all times, but yields a group of features when it is being observed at a given time, either by a scientist like an anthropologist, or by other members of a group who routinely think about their social surroundings in terms of patterns they observe in the behaviors of their consociates and within which they see themselves as being embedded.

If a given culture were always the same, then we would not need to take this level of care in thinking about the nature of culture; in other words, if the observable invariably yields a consistent content, then we do not need to concern ourselves with indeterminacy (Margenau 1949: 297). But the fact is that consecutive observations of cultural patterns can and usually do yield different results and, thus, depending upon the occasion of the observation, the properties or characteristics of a particular culture may be different; as such, the properties of that culture are latent and indeterminate, only appearing as determinate properties in the form of ideas, beliefs, symbols, or actions when there is a specific interpretive action between observer and observed (again, the observer and observed can be either the scientist or any other individual engaged in the act of interpreting her surroundings and interacting with others). In short, the properties of Culture X exist as a potentiality, only to take on definite form at points of interaction among observers (cf. Margenau 1949: 297).

Culture does not simply provide a set of ideas, rules, or concepts that shape behavior; as people organize those variables in their heads culture becomes an observable environment of behaviors that people encounter and interpret and that, most profoundly in children, influences the physiological development of the brain. Neural pathways are formed as we encounter our world and our brains are constructed through constant interactions with the natural and social environments in which we live. In other words, as complex, difficult to pin down, and ultimately indeterminate as culture may be, it is also central to who and what we are as intelligent beings.

Indeed, it is precisely because culture is so important to how humans—and I would argue any intelligent beings[3]—function that despite the problems with defining concepts, we need to at least attempt definition in order to help us use the idea carefully in our discussion. Before doing this, however, it is important to point out that in the 1980s the culture concept was criticized by postmodernist scholars who

[3] Humans are not the only animals on earth who exhibit culture or cultural variation. Chimpanzees make use of rudimentary tools and some cases isolated groups use different tools and techniques for in activities such as gathering ants. See Wexler (2006), p. 184.

argued that it tends to stimulate people, including anthropologists, to think about human societies as static, bounded units that have a considerable amount of internal cohesion and can be clearly identified. This assumption, as Prashad (2006) notes, generates a sense of culture as self-enclosed, and can lead us to disregard the fact that throughout history human societies have overlapped and that boundaries between them, if it is even truly meaningful to talk about boundaries at all, have been extremely permeable.

> Rather than the bounded "thing" that characterized early concepts of culture in anthropology and continues to be a central approach to thinking about culture in many areas of scholarship outside of anthropology and in the popular media, culture is better understood as a process by which people continually contest and reinvent the customs, beliefs, and ideas that they use, collectively, individually, and often strategically, to characterize both their own groups and the groups to which others belong.

Scholars like Pasquinelli (1996) argue that, whether we accept the postmodern critique or not, the use of the culture concept has become so unfocused and diffuse, particularly as a result of its spread to areas outside of anthropology and into public discourse, that it has increasingly come to be seen as "lacking scientific credentials." This problem has caused many anthropologists to distance themselves from the culture concept and to some extent the idea has become marginalized in anthropological circles, where scholars have tended to be more focused on variables such as race, ethnicity, identity, nationalism, and so on.

The problem here is that these other categories, such as race or nationalism, are equally susceptible to being ossified and treated as representing bounded, static groups of people and the more we try to define them, the more their essence seems to slip through our analytical fingers. Anthropologists often go to great lengths to develop nuanced arguments aimed at emphasizing the fuzziness of these categories or noting that people who identify with a category, such as race, may express variable degrees of clarity about the nature and boundaries of a racial category, even while from an analytical perspective race is quite difficult to define. In the end, we are forced into using categories if we want to talk about human behavior, even while recognizing that the categories are inherently problematic. Dropping the culture concept does not get us any further than dropping or focusing on any other category, because all categories we create are necessarily incomplete. Any theoretical category we use will eventually tend towards creating feelings of boundedness about the thing the category attempts to describe or explain and this will generate a problem of obfuscating the nuance, variation, and complexity that exists in the objective world. The category is not the thing it describes, thus no matter what we do it will have limitations and problems. The question really should be not one of whether the culture concept is useful, but one of how can we think about culture in a productive way. This means recognizing that the category of culture is a construct or abstraction used to try to bring together a set of aspects of human behavior—variables—that appear to be evident in the actions and thoughts of many people in a particular group.

Having written the above, there is a further problem in how we use the culture concept that we need to think about. In much of the above discussion, I am looking at how we tend to represent culture as though it is a "thing". In other words, we use it as a specification of a conceptualization about the world. In this sense, we use culture as an ontological commitment to identify something in the world that shapes human behavior and is shaped by that behavior. I find this approach highly problematic because it relies upon a sense of culture as a thing in the world that should be able to be located in some way. This is the way that the concept is used in public discourse as well as in many academic disciplines that work from an uncritical (and unsophisticated) understanding of culture. This is where we get bounded notions of cultures such as Japanese culture or French culture that end up being essentializing constructs, focused on stereotypes, and analytically not very useful.

A second way to think about culture is from an epistemological perspective—culture is a conceptual framework that exists in people's heads and that they use to interpret and understand the world. But how can we use this as a way to think about commonalities in behavior among collectives? This is where the definition needs more precision. *Culture*, from my perspective, *is an indeterminate or latent complex of interpretive variables that exist in individual brains and shape individual behaviors*. This approach to understanding culture emphasizes the epistemological perspective and tends to "locate" culture inside of individual brains. The fact is that the memories—or the knowledge—that is stored as patterns in our brains are idiosyncratic. No one actually truly *shares* anything. Every thought, every memory, and every word we utter is an idiosyncratic product of the unique complex of experiences, memories and ideas that form our lives and form the interpretive variables through which we filter experience of the world.

But the culture concept implies that something is in some way shared by collectives of individuals, and that is often where the problem arises—if it is shared is it also deterministic? To what extent does it shape thoughts and behaviors as opposed to being shaped by thoughts and behaviors? These questions are really at the root of the critique of culture as either a feature of human social organization or as an analytical category. The postmodernists raised the question of how much using the idea of culture tends to *generate* a sense that X people are associated with X culture in some bounded way. In other words, is culture a feature of human groups or is it a product of thinking about humans as groupings of like individuals that limits and essentializes those groups? While I think there is value in contemplating this issue, I also have concluded that it is ultimately a product of confused thinking about culture that fails to differentiate the experiential aspects of culture from the analytical perspective and which, in turn, leads us to think in terms of culture as a determinate, rather than an indeterminate, social variable. And the concept of culture as a determinate "thing" encourages us to think about consistency within an identified collective of individuals, which turns out to be problematic.

Rather than thinking about consistency of thought, ideas, behaviors, etc., we can think in terms of probabilities, or the ratio of the number of incidents of a particular type of event (defined as a thought, idea, action, behavior, etc.) that we find to the total number of observations. In actions like flipping a coin, we can calculate the likelihood that the coin will be either heads or tails on each flip with limited accuracy. We cannot know in advance what the next result will be, but with enough flips—say 5,000 or 10,000, the more the better—we can feel comfortable that the likelihood of getting either heads or tails will be 0.50 (50 %) or that half of the flips will be heads and half will be tails over time. This number represents the limiting relative frequency, which is the fraction of times, over a sufficiently long series of events, in which the outcome of a given event can be represented as something approaching a fixed value. Again, this is does not mean that we will be able to predict with certainty the outcome of any particular flip or string of flips. We might get 20 consecutive flips that are all heads. But over a long enough time, we can, with some degree of confidence, assume that about half the time in a collective of events involving coin tosses the results will be heads (as opposed to 25 % tails and 75 % heads or some other combination) (Von Mises 2011: 359).

Culture can be thought of in a similar way to a limiting relative frequency or the likelihood that, over a sufficiently long series of events, the behavior of individuals in collective X can be represented as something approaching a fixed value. In other words, there will be a degree of consistency in the behavior of most people in collective X, despite the fact that there may be nothing prohibiting variations in that behavior and that, in fact, many variations actually exist. And, of course, the matrices of events, ideas, and behaviors associated with any collective involving human behavior and motivations are vastly more complex than the repeated flipping of a coin. What shapes the outcome of any specific act is the matrix of interpretive variables that exist in each person's head, and because these matrices vary from one person to the next and over time within the same person, we can only think about cultural trends in terms of probabilities. This makes culture inherently indeterminate.

Interpretive variables can be understood as operators that function within the cognitive processes associated with interpreting the world and that generate behaviors. Figure 4.1 shows this as a process in which input in the form of observations/interactions are associated with interpretive variables to form functions that influence behavior/output. These functions should be understood as probability functions, in the sense that when calculating the extent to which we share ideas and knowledge with others we use them to estimate the probability that an outcome associated with

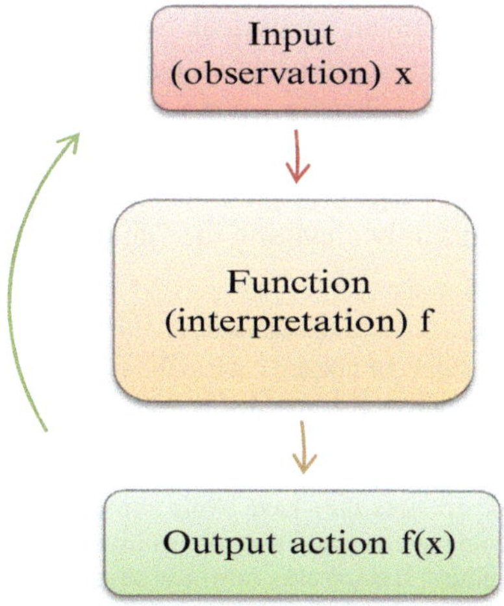

Fig. 4.1 Culture as interpretive process

that variable will be shared with others within our collective. What culture is, if we can think of it as a "thing" at all, is a cognitive domain (in each brain) that forms a matrix field we use to create understandings of social rules and to express and interpret the functional connections that link individuals. This creates a cognitive mechanism in which specific values are ranged into systems of values that we see as being held in common among members of a particular social group. To put this another way, culture is simply the probability that among the individuals of collective y the output of a function f will correspond in some way to an input x. When we think of this group (y) as a collective from either an experiential or an analytical perspective we are conceptualizing culture as "a mass phenomenon or a repetitive event, or, simply, a long sequence of observations for which there are sufficient reasons to believe that the relative frequency of the observed attribute would tend to a fixed limit if the observations were indefinitely continued" (Von Mises 2011: 360). Keep in mind that observations are happening from two perspectives of analysis: (1) that of the agents involved in the collective who are making observations of their surroundings and using those observations to predict likely behaviors of our peers and (2) the perspective of the anthropologist or other researcher engaged in observation and analysis of what he or she observes among the members of a collective.

The first of these is often perceived as being deterministic. In everyday life, we tend to see the social environment as providing a limit—and to the agents involved in that environment it may even appear to be a fixed limit—to behavior that creates the *appearance* that the current conditions of social interaction continue indefinitely. And this is where I think we find culture from an experiential perspective—it rests in the sensibility that the conditions of social interactions in which we operate are consistent in terms of geographical and social space and over some undefined period of time. This is interesting because we often perceive of culture as consistent even while recognizing that things change. Thus, even in scholarly literature we often see representations of something like "Japanese culture" from 1,000 years ago as though it has much of anything to do with what we might call contemporary Japanese culture, a conclusion that is most likely specious.

From an analytical perspective, culture is more productively understood as indeterminate because we know that the observations, and consistent reactions to what we observe, cannot be indefinitely continued. And, in fact, the process of interaction, observation, and interpretation among the members of a collective continually changes the parameters or limits.[4] In other words, simply by interacting, we are constantly changing the "culture" in which we live and operate.

[4] It should be pointed out that the scientific observer (whether anthropologist, astronomer, or chemist) is him/herself embedded in the context and thus is faced with considerably difficulty in separating the experiential from the analytical uses of culture. Anthropologists are at once interested in understanding a particular culture, but are also both shaped by their own cultural experiences and the experiences they have while conducting research. There is no neutral area in which the anthropologist can stand that allows him or her to observe behavior in a way that is not shaped by the cultural frameworks in which he or she is operating and that have influenced his or her own formation as a person.

The key point I want to emphasize here is that culture is indeterminate; it comes into reality as individuals interact with their surroundings, but there is no determinate, concrete thing that we can latch onto and call culture. Culture is not contained in some ephemeral social milieu, but is a complex of continually changing variables operating in the heads of the people who define their own selves in terms of a particular set of contextually shaped constructs (ideas, memories, and behaviors). As a result, culture is idiosyncratic and indeterminate. There is no determinate thing we can call American culture, nor Japanese culture, nor extraterrestrial culture. Instead, there are multiple and varied constructions and interpretations of the social milieu in which people live. Many of these constructions and interpretations are collectivized and, thus, viewed as being shared by the members of any arbitrarily defined social group. The fact that these interpretations to some (variable) extent overlap and generate predictable behaviors and selves that are at some level mutually identified with consistent observable behaviors leads people to think of culture as being "out there," transcending the individual, and having an existence in its own right, governed by its own laws.[5]

4.2 Implications for SETI Research

There are several themes running through the above discussion that have relevance for thinking about the nature of ETI and issues related to how we might communicate if contact is established. First, culture is not distinct from biology—the sensory apparatus individuals use significantly shapes their experience of the world and the manner in which they experience and construct cultural ideas and patterns of behavior. Second, culture represents a context for linking memory, experience, and predictability (past, present, and future) into an interpretive framework that people use to deal with their surroundings. Third, culture is not bounded, nor is it constant,

[5] In addition to avoiding assumptions about one's own culture as collective and "out there" in the world, it is important to recognize that the culture concept can at times be used in ways that "freeze difference" in a manner similar to concepts such as race and ethnicity. The culture concept, when viewed in terms of collectivity rather than as a process that involves some level of collectivization of ideas and behaviors, can become a way of reifying an "other" that is inherently contrasted to the "self", whether it is the "self" of the Western anthropologist or Western society or the "self" of another society as people attempt to contrast, often strategically, their own world with the worlds and ideas of other societies. Abu-Lughod (1991) argues against cultural labels that homogenize the experience of women and, thus, obscure the worlds that women create and within which they interact. This idea can be extended more broadly as a strategy to avoid homogenizing categories such as "Japanese culture" that inherently obscure the variation that exists within the heads of individual Japanese and the complexities of their own interpretations of something or somethings—an abstraction—that is/are represented as Japanese culture by both Japanese and foreign observers of Japanese society. In one sense, there is no such thing as "Japanese culture"; but there is a sense in which many Japanese people, under certain circumstances, *tend* to interpret their surroundings in similar ways and construct their worlds on the basis of assumptions about what is natural and normal behavior.

instead it is in a continual state of change or flux and is indeterminate. Finally, culture is not consistent, it is an abstract amalgam of individual experiences, interpretations, and memories that are treated as though they are consistent but, in fact, involve considerable variation at the individual level.

When thinking about the nature of and the potential for communication with an extraterrestrial intelligence, one can draw a variety of conclusions from this type of understanding of culture. First, it is necessary to give quite a bit of thought to what we *mean* when we speculate on the characteristics of an intelligent extraterrestrial. Most character traits that we apply to humans are heavily laden with cultural values and are not uniform even across our own species. For example, as I mentioned earlier, Drake (2008) argues that any highly organized group of people is going to necessarily require altruism, because the ability to put the needs of the social whole ahead of one's own needs is a prerequisite to creating any organizational structure. On the surface, this seems to make sense; however, when we begin to think about what altruism actually is or means and how it can vary across cultural contexts, the idea becomes complex and opaque.[6]

First off, there is the basic challenge of defining altruism. One way to define altruism is to associate it with virtue, which is a term we usually connect with the character of a person and view as being related to some type of inherent notion of the good. I don't have the space (or the desire) to go into a detailed discussion of the debate between moral relativists and moral objectivists about truth, but the simple fact is that philosophers have never really agreed on whether there is any basic moral truth that would identify what is a virtuous act (see Traphagan 2013). If we align altruism with some notion of the moral good, the question automatically arises—which notion? Different people in different cultures conceptualize good in very distinct ways. So how do we identify an altruistic act if people don't agree on what constitutes a good act?

One way to think about altruism is to view it as involving actions in which an individual does something that helps another individual with some level of apparent risk or cost to him or herself (Jumain et al. 2009). On the surface, natural selection would not seem to be particularly conducive to altruism, because altruistic behavior can work against the reproductive success of the donor. If a 15-year-old pushes someone out of the path of an oncoming bus and gets squashed in the process, he is doing very little to ensure the likelihood of having sex in the future and, thus, reproducing his genes so that his phenotype can be tested against the surrounding environment. Nonetheless, we do see evidence of behavior that appears to meet this definition of altruism among primates, particularly in the form of protection of dependent offspring (Jumain et al. 2009: 142). Other animals also engage in behavior that meets this definition of altruism as sacrifice: Social insects such as ants live in colonies in which workers, who are sisters of the queen, cooperate to raise the sons and daughters of the queen, squirrels engage in alarm calls to warn kin of approaching danger, and the offspring of scrub jays may remain in the nest to help parents raise additional offspring.

[6]For an interesting and useful group of discussions related to altruism and extraterrestrials, see Vakoch (2014b).

Perhaps the most dramatic example of this type of selfless behavior in nature comes in the form of sexual suicide among drone bees. Honeybee sex occurs in mid-air when drones swarm the queen and compete for chances to mate. The big moment comes when one drone dives in, grasps the queen, and inserts his sexual organ into the queen's reproductive tract. When he ejaculates, the explosive force ruptures his sexual organ and he drops off of the queen and dies, adding new meaning to the comment, "was it good for you?" The queen mates with several drones in this way, leaving behind a trail of dead, selfless bees who gave their lives in pursuit of the ultimate reward—passing on their genes (Forsyth 2001: 42, note that not all bee species do this).

A question arises, of course: Is this really altruism? Doesn't altruism imply some form of intent on the part of the donor? Do drone bees think to themselves as they circle the queen: "I'm going to spread my genes for future generations even if it kills me!" I've already established that we can't really know what is going on in the heads of other organisms, but I'm going to go out on a limb and suggest that bees probably don't have this running through their brains when they mate. If we work from a strictly biological perspective, then, yes, bees can engage in a sort of altruistic behavior. In fact, much, although not all, of what is termed altruistic behavior in animals can be explained through the theory of kin selection, which suggests that altruism may be genetically encoded in humans and other species and operates as an evolutionary mechanism that selects for those behaviors that increase *inclusive fitness* which is a measure of how the actions of one individual (donor) affects the fitness of others (recipient) (Wenseleers et al. 2010: 138). When looked at in terms of kin selection inclusive fitness refers to the evolutionary success of an organism being evaluated on the basis of the number of offspring it is able to produce and how those offspring themselves succeed (see Hamilton 1964). In other words, an organism can improve its overall genetic success or transmission of its genes by altruistic social behavior in which it may be beneficial to sacrifice oneself in some way to ensure that one's genes get passed on and have a chance to be reproduced by one's offspring. If this is right, then there is hope for Drake's hope that ETI will be altruistic, at least in one sense of the word.

But this is not altruism in the sense of intentionally sacrificing oneself without regard to any personal gain that might occur, and this is the type of altruism that Drake associates with intelligent beings. Although Drake recognizes that altruism can vary from one society to another—as he puts it "one world's altruism may be another world's barbarism"—he, and most others who argue in favor of altruistic extraterrestrials, think that altruism in the form of beneficence must at some level be a universal trait for technologically (and socially) advanced intelligent species. There are, of course, problems with this formulation, the most notable of which is the recognition that group and individual needs are not necessarily at odds and, thus, individuals may *only* align themselves with the needs of the group when those needs correspond with their individual interests.[7] I may align my own interests with those of administrators in my university simply because they write my paycheck, not

[7] For a fascinating exploration of altruism from both biological and cultural perspectives, see Oakley et al. (2012).

because I necessarily agree with their decisions about university policies and practices. In other words, being "altruistic" may simply be a form of selfish behavior. In fact, in a recent class on biomedical ethics that I taught, both my students and my teaching assistant were fairly adamant that altruism doesn't even exist. The argument in class was that at some level *all* other-centered behavior is ultimately self-serving. Even when one saves another person at personal sacrifice the act may well be motivated by selfish desires. This is the so-called martyr complex, in which people sacrifice themselves for the purpose of being perceived as heroic.

Thus, when we ask questions like, "will ET be altruistic?" or "will ET be bent upon eradicating inferior beings from the galaxy," we are skipping the more important question of whether or not altruism is a universal trait of intelligent beings or whether, even if it is, it would manifest itself in the same way among different types of intelligent beings. In fact, when it comes to thinking about ETI, the question of whether or not ETI will be altruistic, nasty, or anything else is clouded by the ways in which we construct the culture of ETI in our own imaginary and make assumptions about the extent to which there is a reasonable possibility that extraterrestrials will be diverse, just like we are. Questions and debates about the likely character of an alien civilization assume that aliens will be quite uniform in the way they construct their own imaginaries and react both to each other and how they are likely to react to contact with humans (or anyone else). Reflection on our own case, and the above discussion of culture and cognition, makes it clear that if they are anything like us, this is unlikely to be the case.

4.3 The Star Trek Imaginary

Indeed, as I discussed in Chap. 3, a tacit assumption of much of the literature dealing with contact with extraterrestrial intelligence is the idea that an alien civilization will be culturally unified, *unlike* our own world. The idea seems to be that technological and social progress inevitably leads to greater levels of unified organizational structure associated with higher levels of moral unity and overall niceness, but this is an assumption of human, and particularly Western, perspectives that have a teleological notion of cultural evolution in which there is a universal outcome to processes of cultural change. Advanced, in this formulation, becomes inexorably associated with culturally, morally, and politically unified.[8]

[8] One good example of this from the corpus of work on SETI can be found in Dick's interesting paper (2008), "The Postbiological Universe" in which he works from the assumption that a "central goal" of cultural evolution is increasing intelligence. The idea that cultures necessarily evolve, rather than change, is based upon Western (cultural) notions about the nature of human social organization in which certain social structures are more advanced than others and that there is a directionality—that implies improvement—to the flow of cultural change.

In essence, this is a very (naïve) anthropocentric approach and tends to assume an underlying likeness in any intelligent being and in the ways in which "culture" is expressed and shapes civilization formation among all intelligent beings. However, if we do look at our own world as an analog for what we might find elsewhere, we are faced with the fact that rather than a single, unified, culture, it is quite possible that we will be dealing with a world that is fragmented into different cultural frameworks, much like our own, consisting of people who may not respond to contact with us in a uniform way. As I discussed earlier in this chapter, technological advancement on Earth has not necessarily been associated with increased political and social integration—think World Wars I and II. In fact, one of the trends of the later twentieth and early twenty-first centuries has been increased political *fragmentation* with the break-up of the Soviet Union and other moves among ethnic and cultural subdivisions within larger political structures to seek increased autonomy and independence—think Iraq, Spain, Chechnya, the Ukraine, etc. This list is really quite long.

This teleological speculation about ETI is based upon what I call the Star Trek imaginary, in which people assume that humans are headed toward either unity or destruction, thus any advanced civilization will have found unity, because if they hadn't they would have blown themselves up. This might be true, but there are other alternatives. It may well be that civilization is normally cyclical with periodic rises and falls in both technology and in levels of unification—think Roman Empire, British Empire, American Empire, etc. To date, nothing in human experience suggests that we will actually break out of that cycle; I certainly hope we do, but the evidence just isn't there to be terribly optimistic. Any extraterrestrial civilization we might encounter may well have risen and fallen many times in its history and we may just happen to hit it at the moment of a high (or a high that happened 1,000 years ago if it's 1,000 light years away)—but the elements leading to collapse may still be lurking. And even if the argument that any extraterrestrial civilization we might encounter will be much older than ours by virtue of the time needed for us to intercept a transmission is right, there is no reason to believe based upon the evidence we have that said civilization has not blown itself up and risen from the ashes many times throughout its history. In fact, if that civilization is multi-planetary, this seems quite possible.

Even if the experience of our planet is dissimilar to that of another world, it seems reasonable to think that we will be dealing with beings who are shaped by common memories (among themselves) and who will share, but who will also debate and contest, ideas developed within the frameworks of those common memories and experiences about what to do with the fact of having contacted humans. This problem is exacerbated when we take into account the high likelihood that alien beings may have sensory organs that are different from our own and, thus, may process experience and translate that experience into cultural frameworks in a way different from our own. And even if such experiences and memories can be seen as "common" they must be understood in the manner identified above as being highly particularistic, based upon individual experience—unless, of course, we encounter an alien society in which individual beings are cognitively unified in some way and, thus, actually do share a single experience of the world. This is the type of ETI that we might encounter if Dick's (2008) notion of a postbiological intelligence being our likely contact is correct. In that situation, the meaning of culture for such a being becomes extremely difficult for humans to understand or even imagine.

4.4 Symbols and Meaning

One solution to this problem that has been proposed is that members of both societies will share the capacity for symbolic interaction. In order for the transfer of information to occur, intelligent beings need to be able to make one thing stand for another thing. If humans didn't have the capacity to make the color red stand for the command to stop, we would have a very difficult time transferring the information necessary to make driving reasonably safe most of the time. However, in much of the symbolic activity of humans, the linkage between thing and thing signified is arbitrary—there is no necessary correlation between the color red and the meaning "Stop!" Vakoch (2010) argues that the use of icons, or signs that physically resemble the thing signified may provide a way out of the problem of the arbitrary linkage between sign and symbol, even while there remains the fundamental problem that the interpreter may still not understand the physical correlation between icon and the thing it signifies. Indeed, if an extraterrestrial has different sensory organs from humans the idea of what constitutes iconic symbols may not work—would an image of Abraham Lincoln on a five dollar bill look like the actual Abraham Lincoln to a creature that uses echolocation to sense its surroundings?

Perhaps part of the way to deal with this is to recognize that the point of contact will represent a context in which not only is the intended message interpretable, but the method of communication and the method of representing information is also interpretable—and perhaps more so than the intended meaning. What I would like to suggest here is that should we encounter evidence of extraterrestrial intelligence in the form of a signal (directed at us or not), we should not only be concerned with deciphering the meaning of the signal in terms of intended content, but also in terms of what it tells us about the thing that sent it. This is one way to try to limit the extent of anthropomorphic assumptions about the nature and communicative approach of ETI. A signal is not only a conveyor of explicit information, but also of implicit information about the sender. The fact, for example, that humans have been sending television images out into the galaxy for several decades could tell extraterrestrials a great deal about us, if they are able to recognize that those signals contain information that can be represented in a visual medium. The simple fact that we send out electromagnetic signals that can be interpreted visually and aurally indicates a great deal about how we process the world—that we have sense organs that translate sound and light into information we can use, as well as the types of limitations (in terms of the light spectrum, for example) of human senses. An alien intelligence that recognizes this fact would then have a basis upon which to create a message that we might be able to understand.

The messages that have been sent out to date, such as Drake's approach of broadcasting pictures and binary information that requires no prior understanding of our technology, are an attempt to anticipate the capacities of another civilization of intelligent beings. Drake's message, for example, provides some basic information about us and our knowledge, including numbers from one to ten, the human form, DNA structure, hydrogen and carbon atoms, and information about our solar system. Drake himself has noted that when he presented the message to different scientists, they had trouble interpreting the entire message—instead only being able to understand the sections immediately relevant to their own area of expertise—leaving us to wonder how well extraterrestrials would do if they stumbled across the message.[9]

Indeed, the difficulty in interpreting the intended meaning of the message suggests that another approach might well be taken. Instead of primarily being concerned with the content of a message, we might want to consider being concerned with what the message tells us about who sent it. In Drake's message, there are several subtexts that convey information about us that are not part of the intended meaning. For example, the manner in which the message is constructed would suggest that we think in terms of binary relationships—we encode information in terms of 1's and 0's—and understand two-dimensional images. Extraterrestrials might assume that this is how our language works or represents how humans organize thought in general, an assumption that would be misleading. However, the fact that the message represents information in a visual manner, like our television signals, would suggest, correctly, that we are visually oriented beings. If the message was interpreted as being sent by "an alien civilization" for the purpose of making contact, then it would suggest quite inaccurately that we are a unified society or culture interested in communication with civilizations in other parts of the universe. As I argued in Chap. 1, it is quite possible that for the majority of humans contact with ETI will not be particularly meaningful.

In this chapter, I've tried to show how the ability to both imagine and communicate with ETI is intertwined with a combination of our cultural and biological being as humans. As we speculate on the nature of ETI, we need to focus on the questions: "What are the implicit indicators and forms of information about ETI and ourselves that are contained in any message sent or received?" "What are the assumptions about the nature of culture, civilization, and intelligence that we bring to the analytical table when thinking about ETI? These questions should lead to consideration of how we might develop useful tools to interpret implicit information in any message we might receive and also how to encode that type of information in any signal we might send. In many respects, focus on how to interpret implicit information may be more important than how to interpret the explicit message, given the potential differences in culture and biology that might exist between ourselves and an extraterrestrial other, as well as the inevitable differences in personal intentions and interpretations that will be fundamental parts of contact on either side. It is to this issue—how we might interpret and respond to contact with ETI and how culture may influence that process—to which I turn in the next chapter.

---

[9] Interview with Frank Drake, Astrobiology Magazine, 27 August 2007. http://astrobio.net/news/modules.php?op=modload&name=News&file=article&sid=2441&mode=thread&order=0&thold=0

Chapter 5
Knowledge Production in the Encounter with Alien Others

The invention [of culture] is "controlled" by the image of reality and the creator's lack of awareness that he is creating.

— Roy Wagner (1981: 12)

Throughout much of its history anthropology has explicitly focused its intellectual gaze upon the understanding of seemingly "alien" others whose languages, beliefs, patterns of living, and social structures were often viewed as remote from the societies of the industrial West—England, France, Germany, and the US—in which the discipline developed. In the formative years of anthropology, ethnographers normally couldn't be in direct contact with the extra-societal others who were the object of their studies. Indeed, early, "armchair" anthropologists of the nineteenth century, such as James Frazer, E. B. Tylor, and Lewis Henry Morgan, (although Morgan did also conduct direct data collection among the Iroquois in addition to the armchair variety of research) worked under conditions not entirely unlike those of SETI researchers today. Limitations in technology (specifically transportation and communications technologies) dramatically restricted the types of interaction available for social scientists interested in contacting and understanding a distant other. Communication was slow, requiring weeks or months to request and then receive information from individuals (often missionaries) living in distant places to be analyzed by anthropologists living in locales such as the US or England. When data were eventually received, such as the kinship data collected from numerous parts of the world by Morgan in the mid nineteenth century (cf. Trautmann 1988), interpretation was largely based upon theoretical frameworks and assumptions that had a decidedly Western tinge—specifically social Darwinism and cultural evolution of the Spencerian variety I discussed earlier—and exhibited overtly teleological underpinnings associated with progress (cf. Spencer 1969). Interpretations of data were difficult to test through what would later become true ethnographic research in

J. Traphagan, *Extraterrestrial Intelligence and Human Imagination*,
Space and Society, DOI 10.1007/978-3-319-10551-2_5

the form of direct contact and long-term participant observation. Although the nineteenth century style of armchair anthropology was replaced by ethnographic fieldwork in the early twentieth century, instances of anthropology at a distance continued to occur.

A great deal of ink has been spilled in the SETI community contemplating not only what ETI might be like, but also the significance of contact with an extraterrestrial intelligence for human society. Conventional wisdom among SETI scholars is that contact with ETI will be a major moment in human history; it might even be a game changer for the future of humanity as it deeply influences how we think about and understand our place in the universe. Some recent work has even tried to develop methods for quantitatively gauging the significance of contact with ETI. Almár and Tarter (2009), for example, modify the Tarino scale for judging the potential social significance of asteroid impacts as a model for constructing a scale to judge the potential significance of an announcement that extraterrestrial intelligence has been discovered. The authors develop a table of assigned values related to variables such as whether the message is targeted specifically at Earth, how far away the sender is, they type of discovery (e.g. a steady signal or something found in archival data). This is an interesting idea, but these types of things are extremely difficult to quantify, as any social scientist would tell you.

Let's take as an example the question of the type of phenomenon. The authors assign a value of 6 for earth-specific messages, 5 for omnidirectional messages, 4 for earth-specific beacons, 3 for omnidirectional beacons, 2 for leakage radiation, and 1 for traces of astroengineering activity. The higher the number, the more significant. This seems to make sense on the surface. If ET is aware of us and sending a message aimed at us, this seems quite a bit more significant for us (and potentially threatening) than if we just happen to stumble upon leakage radiation from ET that shows no awareness of our existence. The problem here is that the numbers are completely arbitrary, because we don't know how people will construct the differences between, say, 5 and 6 in their own heads. Different people are going to construct the relative significance of these events in very different ways and how they do this will be influenced by their cultural context.

Social scientists have long been aware of this problem, as it occurs frequently in surveys when people rank likes and dislikes. When asked to assess relative likes of ice cream on a scale of 1–6, I might put chocolate at a level 6 and vanilla at a 1, depending on what other flavors are possible choices. In fact, I like most flavors of ice cream, so the fact that chocolate is at the top and vanilla at the bottom really doesn't tell anyone much about the relative value of the two—I like both of them a lot. If squid ink flavored ice cream were on the list (I've eaten it), it would be a 1 and vanilla would be a 2. But *the difference between 1 and 2 would be vastly larger than the difference between 2 and 6 (chocolate) in my mind.* Squid ink ice cream tastes really bad to me, so it's really off the scale when compared to vanilla or chocolate.

Scales like this when applied to human emotional responses usually significantly over-simplify what is really going on in people's heads—to the point that the data may not really be very meaningful from an analytical perspective. In fact, human emotional responses are very difficult to measure in a quantitative way and the development of measures normally requires a fair amount of qualitative research to determine what kinds of questions need to be asked and how to ask them before developing any sort of quantitative instruments. The authors suggest that the standing committee on SETI of the International Academy of Astronautics use this as a way to assign a value of any announcement so that they can at some level gauge the impact. Well, the same problem will arise when they go over the relative levels in their own heads that happen when I think about ice cream or the general public thinks about the significance of contact with ET. What this measure is much more likely to do than anything else is over-simplify discussions about what the impact might be, because it over-simplifies human behavior, the complex interactions between individuals and their social and cultural contexts, and the importance of values in generating human responses to important events. In short, the Rio Scale doesn't represent good science.

Indeed, the Rio Scale, as Almár and Tarter call it, is as much a product of assumptions that contact with ETI *should* be significant for humanity as a whole, rather than being a scale that will in any way accurately assess how people might really respond. This is a value judgment that influences the entire construction of the scale (and the article in which it is proposed). I mentioned in Chap. 1 the simple fact that there is a distinct possibility that while astronomers, politicians in industrial societies, and anthropologists may see learning of the existence of ETI as a huge event, most of Earth's population probably won't care much—they will be too busy trying to survive. As I write this chapter enjoying my comfortable summer in Tokyo, I watch the news on CNN that, at the end of 2013, over 51 million people in the world were displaced from their homes by violence. It's the highest level of people displaced by violent conflict since the end of World War II. What significance score will be appropriate for these people and how would that score differ from those sitting in nice apartments in Tokyo or Boston, people who have time to think about and worry about contact with ETI? Will it really matter to these suffering individuals and the millions more who live in abject poverty that a few scientists in San Jose found evidence for the existence of intelligent life on another world?

I raise this not to question the value of SETI—in fact, I think it is a highly worthwhile endeavor and it is sad that quite a bit of the funding for SETI has dried up in recent years. But I think we also need to caution against naïve assumptions about the potential influence and significance of SETI for human history. Contact with ETI *could* be a game changer for humans. Perhaps it would get our politicians to begin to truly act on solving some of the world's many problems, or maybe humans would feel unified in the face of a perceived common threat. Who knows? But I'm not going to hold my breath for these things to happen if contact occurs.

There are other scenarios that also should be considered, because regardless of the influence, rather than a simple, unambiguous act of discovery, initial contact with extraterrestrial intelligence will become a context through which a new imaginary is generated and understood as knowledge is produced; in all likelihood the context of initial contact will be formed on the basis of very limited data and, inevitably, interpreted through the lenses of our own cultures and the theoretical frameworks that are in vogue among intellectuals and others at the time contact occurs—similar to what happened in the early days of anthropology. As such, it is unlikely that there will be a single meaning that humans draw from contact with ETI. Meaning will be varied, complex, and inevitably a source for generating conflict. And although the negotiation of that meaning will contribute to the creation of an imaginary in which humans are not alone, one wonders how different that imaginary might be from the one that developed after Lowell's "discovery" of intelligent life on Mars. We should not lose sight of the fact that in a sense, discovery of an extraterrestrial civilization has already happened once on Earth, despite the fact that Lowell's big moment for humanity turned out to be a flop—and a broad imaginary that evolved into science fiction among other things emerged from that flop.

In order to think about the question of how humans, and scholars in particular, might react following contact with ETI, we can look into the history of anthropology and explore the "anthropology at a distance" used in the discipline's formative years and to a lesser extent later on as a result of political limitations that prevented long-term fieldwork in some instances. I want to focus specifically on the work of anthropologist Ruth Benedict during World War II as an example of the complex interplay of assumptions, data, and misinterpretations that might form the basis on which knowledge about and an understanding of an alien civilization could unfold.

Before looking at Benedict's work, however, let's begin with a brief review of anthropological thinking about culture, as a reminder of the ideas explored in Chap. 4. Anthropologist Marvin Harris has noted that the beginnings of anthropology are to be found in the inspiration of the natural sciences and scientific method; early anthropologists worked from the assumption that sociocultural processes are governed by "lawful principles" that can be understood in terms of causality and that are discoverable by an objective observer (Harris 1968: 1). This mindset led to a tendency among anthropologists, and many others, to represent distinctly defined cultures as relatively bounded entities and to posit culture as having deterministic qualities, significantly shaping the behaviors and thought patterns of the people who live in a particular context.

The second half of the twentieth century, in particular, was characterized by much more nuanced work where scholars like Wagner (1981) and many others since have viewed culture as fluid and permeable and conceptualize it as a process of invention in which particular "cultures" arise out of an intersubjective dialectic between the individual and his or her social environment—an approach that is reflected in the tentative definition of culture I offered in the previous chapter. The anthropologist is not remote from this process, but, instead, can become intimately involved in the invention of a particular culture—understood not so much as a group of people, but as both an analytical category and a popular framing of a particular group of people—through translating, interpreting, and writing about what she observes, as well as through the daily interaction associated with the activity of fieldwork.

5.1 Ruth Benedict and the Invention of Japanese Culture

Nowhere, perhaps, is this more clearly evident than in the research Ruth Benedict (1946) conducted during World War II on Japan. To begin, Benedict's work is, if not exactly armchair anthropology, a decidedly good example of "anthropology at a distance." Benedict was commissioned to conduct her study of Japanese culture during the mid 1940s by the US government. The object was to provide a report that would explain and predict what seemed to be very alien[1] Japanese behavior and that could be used during what was, by 1944, the anticipated invasion of Japan. In other words, her work was to be an important explanatory guide in the project of invasion and eventual social engineering that would become the Occupation of Japan.

If one thinks for a moment, this was not an easy assignment—anthropologists by the 1940s conducted research through participant observation and long-term residence with their informants, a methodology known as ethnographic fieldwork. In order to do our job, anthropologists really need to be in close contact with those we want to learn about for fairly long periods of time so that we can talk with them and experience life in a way that approximates their own experiences, as well as challenging our own assumptions about what is normal behavior for humans. Benedict couldn't really hop on a plane in the middle of WWII and drop in on Japan, asking her potential informants, "Do you mind if I study your culture so that after we beat you in this war, we have some good ideas about how to govern you?" What is not particularly well understood among the general public, and even some non-specialist scholars with interests in Japanese culture, is how Benedict's research was actually carried out so that she could compensate for this significant methodological limitation.

Benedict did not actually conduct a study of Japanese culture or society through traditional ethnographic methods of direct participant observation and long-term co-residence with Japanese people; instead, due to the war, she was forced to turn to what appeared to be the next best thing—Americans of Japanese descent who were confined to internment camps in the desert Southwest. Obviously, in retrospect, this should raise red flags about Benedict's study. As Ikegami (2003: 1370) points out, Benedict's research subjects were "passive and cautious in their replies to her questions" faced with an authority figure representing the very government that had forcibly removed them from their homes and placed them in the camps.

Interestingly, this issue was not widely discussed by the scholars who reviewed Benedict's book shortly after its publication in 1946; on occasion, the problem was noted, but for the most part it was overlooked. It is only much more recently, such as in Sonya Ryang's work (2004), that open discussion has ensued about how Benedict's research contains flawed conclusions in part because the conditions of her data collection were limited by her inability to actually travel to and make direct contact with individuals within the Japanese context.

For my purposes here, it is not important to go into details about the empirical and interpretive errors that exist in the *Chrysanthemum and the Sword.* Ryang (2004: 33) notes many of the problems and specifically points out the fact that Benedict's development of linguistic data from Japanese is not supported by either sociological or historical data—she tends to select words from her informants and

[1] It's important to recognize that Benedict herself was not trying to represent the Japanese as incomprehensibly alien. Rather, her primary aim was to show that if we understood the cultural logic at the foundation of Japanese society, we could understand the behaviors and their motivations that seemed so alien to Americans during the war.

from literature without contextualizing the terms and understanding how they are conceptually used by Japanese, and in her work these terms tend to become keywords for representing and understanding Japanese culture and behavior. More important than the specific errors in Benedict's research is that fact that she takes a totalizing approach to understanding Japan—specific words, ideas, or concepts evident in sources such as Japanese literature become used to broadly explain, often in a single brushstroke, all or most elements of Japanese behavior. In part this is a consequence of the theoretical framework that Benedict uses, as well as the general lack of detailed empirical data about Japan from which she could draw conclusions—this last point being very important for our discussion here.

While the flaws in her work are significant, perhaps more important is the fact that *The Chrysanthemum and the Sword* becomes, as Ryang (2004: 48) notes, "paradigmatic," playing "a crucial role in the postwar social science discourse on Japan" both among native Japanese and Western scholars. In fact, so thorough was the assumption that Benedict had accurately described and explained Japanese culture as it really is (was), only very rarely was it noted that her work had not actually focused on Japanese people at all. Many of the attitudes toward *The Chrysanthemum and the Sword* at the time of its publication are summed up nicely in a review written by University of Hawaii and later Yale anthropology professor John Embree, the first anthropologist to produce an ethnographic study of Japan, which was published just before the war in 1939 (1964). In his 1947 review of *The Chrysanthemum and the Sword* he states. "…Dr. Benedict, with the soft words of a fox spirit, leads the reader into the forest of Japan and before he knows it she has him bewitched into believing that he understands and is familiar with every root and branch of Japanese culture" (Embree 1947a: 11).

When I first read this I thought he was being sarcastic, but throughout the review Embree's only real criticism of Benedict's book is that she fails to recognize that Japan is an old culture while the US is a new one—itself a rather dubious comment since Japan underwent a radical social transformation in the second half of the nineteenth century.[2] He goes on to state, "[t]he frontiersman and the nomad are more likely to be individualistic braggarts than is the village bound peasant who must face his same neighbor day after day…A man of an old peasant culture such as the Japanese is likely to be more meticulous in his etiquette and sense of reciprocal duty…" (Embree 1947a: 11). Leaving the rather romanticized and totalizing language aside, we see that Embree believes Benedict allowed us to gain entrance into an almost impenetrable cultural "forest" vastly different from ours because it was a "peasant" society based upon village social organization—I'm not quite sure where the thriving metropolises of Tokyo or Osaka fit into that peasant forest—as opposed

[2] In one sense, Embree is correct that Japan has a much longer history than the US, but the US Constitution has been in force throughout a period in which Japan has undergone two radical transformations of it social, political, and cultural fabric—the Meiji Restoration of 1868 and the following industrialization of Japanese society and the US Occupation with its associated political and social changes, the second of which was happening while Embree was writing. This raises questions about the meaningfulness of describing Japan as old and the US as new.

to urban, American, individualistic social organization. He does not pay much attention to the fact that her research wasn't conducted in Japan—indeed, he takes it for granted that Benedict wrote a book about Japanese culture rather than Japanese American culture.[3]

Interestingly enough, even in another review that Embree wrote on the *Chrysanthemum and the Sword* in which he compares it to a book on Japanese living in Hawaii, he takes for granted the idea that Benedict did, in fact, explicate Japanese cultural patterns and behaviors through her research (Embree 1947b). Indeed, Benedict, although openly identifying her fieldwork locale in the US, ultimately represents her research as being about Japanese people and Japanese cultural values. And, as is evident in Embree's review and those of other scholars at the time, this idea went largely uncontested (cf. Clyde 1947; Embree 1947a, b; Raglan 1948; Ackerknecht 1947), although Morris in his 1947 review points out that "in normal circumstances no one would think of writing a serious book without first spending a considerable time observing at first hand the actual behavior of the people concerned" (Morris 1947: 209). Morris quickly puts this problem aside and lauds *The Chrysanthemum and the Sword* as "the most important contemporary book yet written on Japan. Here, for the first time, is a serious attempt to explain why the Japanese behave the way they do" (Morris 1947: 208). Apparently, the lack of normal circumstances provided the justification to ignore his rather important point about what is necessary for an ethnographic work to constitute reliable and valid research.

What is important to gain from these examples is that the *Chrysanthemum and the Sword* became, as noted above, the cornerstone of the ethnographic, and non-ethnographic, corpus of scholarship on Japan, despite the lack of quality empirical data upon which Benedict's interpretations are based. Benedict's contemporaries and many who followed took her work at face value and accepted the idea that she had, in fact, produced a study of Japanese culture. The problems that lurked within the realities of having to do "anthropology at a distance" were overlooked by Benedict's colleagues and many of those who became interested in Japanese culture, and instead her book became the paradigmatic cornerstone of Japanese studies and shaped the major questions posed and studies produced by "Japanologists", most notably the focus on the Japanese psyche or personality (self) which dominated research on Japan into the 1990s (Ryang 2004). Although anthropologists began consciously moving away from Benedict's construction of Japanese culture by the 1960s, her work has been cited hundreds of times and continues to be cited particularly in cross-cultural psychological studies—not as a book about Japanese Americans during World War II, but as a book about Japanese people and their culture particularly when it comes to discussions of shame vs. guilt oriented moral systems.[4] In fact, a quick search on Google

---

[3] It's worth noting that Embree backed away from his support for Benedict's work a few years later, shortly before his death in an automobile accident in 1950. See Ryang (2004).

[4] Perhaps most striking is that when the book is cited today, its conclusions are often presented as constants of Japanese culture, impervious to historical circumstances, such as the influence of American concepts of individualism that became common during and after the Occupation, despite the fact that the book was published more than 60 years ago.

Scholar and the Social Science Citation Index shows the book as having been cited 3,311 times, and in 2014 it has been cited several times in psychology journals as I write this chapter in mid-July.

In essence, the publication of *The Chrysanthemum and the Sword* initiated a process by which Japanese culture was invented as a Western, and particularly American, imaginary. Benedict's work set out the parameters of what would be considered the basic elements and core values associated with Japanese culture and the Japanese psyche for years to follow, and a great deal of the research produced supported the conclusions, either directly or indirectly, that Benedict had drawn through her anthropological research at a distance. As people read and followed her work with further research, analysis, and publication, a sense of Japanese culture and Japanese behavior being represented accurately—and completely—in the basic ideas put forth by Benedict prevailed among scholars working in areas such as cross-cultural psychology and cross-cultural communication studies, as well as in the broader community of non-scholars who were simply interested in Japan. In short, Benedict's at-a-distance take on Japan, became Japan itself for many, and perhaps the majority, of Americans throughout most of the second half of the twentieth century. The influence of this cannot be overstated—Benedict's work was central in the US government's approach to re-organizing and engineering Japanese society following the War and was widely read by an American public interested in understanding the enemy they had just conquered and whose country they were now occupying but found in many ways mystifying.

However, what was being created was not a "true" understanding of Japan—if such an understanding of any culture is actually possible. Rather, what emerged in the discourse on Japan was an imagined culture based upon values and psychological orientations—with an emphasis on the concept of shame—that seemed important to Benedict. Indeed, the book is an application of theoretical ideas she developed in an earlier work, *Patterns of Culture* (Benedict 1934), in which she used psychological idioms (although not Freudian in nature) as a means of creating configurations or categories of cultural types that, in turn, were imprinted in the minds of those living in a particular cultural milieu. In other words, Benedict's understanding of how culture works and what culture is should be seen as a direct result of the academic context—with its considerable interest in psychology—in which she was trained at Columbia during the 1920s and that continued to be influential as her career developed.

5.2 Implications of Anthropology at a Distance for SETI

So what does all this stuff about Japan have to do with SETI? Oddly enough, this foray into the history of the anthropology of Japan is quite relevant to our discussion, because it gives us an empirical basis for thinking about how an encounter with an extraterrestrial technological civilization might play out here on Earth and the extent to which our imaginary might be constructed following that encounter.

The conditions of research under which Benedict was forced to operate prevented a strong dependence upon gathering quality empirical data and eliminated the possibility of conducting true ethnographic research in the form of participant observation and long-term fieldwork. Indeed, relatively little of the data she used were actually collected by her—instead she borrowed data collected among Japanese Americans by psychological anthropologist Geoffrey Gorer during the War in the relocation camps and also used the work of another psychological anthropologist, Weston La Barre (Gorer 1943; La Barre 1945), neither of whom were trained as Japan scholars, although she did also collect some interview data of her own (Ryang 2004: 17).

As noted above, given the lack of valid empirical data and the very limited scholarly resources available on Japanese culture and behavior (cf. Ryang 2004: 16), as well as her general tendency toward emphasizing (psychologically oriented) theory over data, Benedict essentially took the little data she had and worked it into the theoretical framework she had developed earlier in her career in her book *Patterns of Culture* (1934) that categorized Native American cultures on the basis of personality traits associated with a particular group of people. Benedict's study of Japan—from afar—set in motion a conceptualization of Japan and the Japanese that has influenced scholarship and policy-making related to that society right through until the present. And a great deal of what she wrote has turned out to be either a very simplistic representation/explanation of Japanese culture or fundamentally inaccurate—yet her work continues to be influential.

If we turn to a bit of speculation about our initial encounter with an extraterrestrial intelligence, it is not difficult to imagine an analogous process occurring should contact be made. The first scientists to encounter a signal from an extraterrestrial intelligence will likely have a very limited amount of data from which to work. If we simply capture a signal sent out that is not directly aimed at attracting the attention of an alien civilization (such as our own), then the data may be extremely difficult to understand in such a way that we can develop a clear sense of what we are looking at in terms of meaning or in terms of what the message might imply about the nature of ET. Maybe our first step will be to try to gauge the significance of our new knowledge about our place in the universe using Almár and Tarter's Rio Scale with all of its problematic assumptions about human behavior and over-generalized notions about the ways in which humans interpret their surroundings.

Furthermore, the problems we face with a specific message are not simply a matter of translation; even if we can figure out specific meanings of linguistic constructs that correspond to something in our own language, we will have no cultural framework with which to interpret how those meanings apply to an alien society. In the case of Benedict, even when she had the base-level fact that she was dealing with another human society that had the same fundamental structures (albeit different in their manifestations) as American society—systems associated with religion, kinship, government, etc.—a lack of sufficient data and an unconscious and to some extent conscious tendency to fit an alien culture into a framework that makes sense to an American mind led to a casting of Japanese culture along particular lines, many of which were not accurate.

The odds are that, without much data and, thus, any real understanding of an extraterrestrial culture with which we make contact, we will interpret what we find in terms of values, structures, and patterns of behavior associated with *our own* culture. So obviously, the solution to this problem is that we need to be very careful not to uncritically apply human value systems as we try to unravel anything we might get from our extraterrestrial interlocutors. The lesson from the history of anthropology is that we need to be very careful to avoid ethnocentric, and in this case anthropocentric, application of human value systems in our analysis and attempt to understand the meanings of any message we might receive from ETI. That seems obvious and we've learned a lot since Benedict and developed much more sophisticated techniques of data collection and interpretation, as well as complex ideas about the nature of culture. We're cool.

Wait a minute. How do we prevent the values of earthly civilizations from tainting our understanding of the civilizations of ETI? What Earth culture do we need to be careful about holding at bay to prevent anthropocentric analyses? Do we have to watch out for how ethno/anthropocentric American attitudes influence our interpretations? Ethno/anthropocentric Japanese attitudes? Ethno/anthropocentric English attitudes? Ethno/anthropocentric Zambian attitudes? You get the picture. Of course, this is problematic because there is no, single, human culture (or civilization) on Earth, thus there will be multiple avenues for interpreting the meaning of whatever we might decipher out of a transmission from ET. Humans have no single framework for understanding the world, thus we aren't going to have a single interpretation of what a signal's intended meaning might be or what its impact might be for humans, even if we can decode the basic content of the transmission.

Furthermore, as many differences as we can find among human cultures in terms of how we perceive of, interpret, and categorize our surroundings, it is reasonable to assume that a truly alien society would consist of beings who do culture in ways distinct from those of humans (Vakoch 2007). The capacity to *do* culture in a relatively consistent way among human beings—even with all of the differences we find in specifically how cultural values are constructed and expressed—is heavily dependent upon a common set of sense organs, as I discussed in Chap. 4. And even with this, as noted earlier, neurological studies have shown that differential experiences and forms of stimulation during developmental processes shape the connections among neurons and thus influence the construction of the neural networks that are basic to human behavior and thought (Wexler 2006). What would culture look like when applied to a being with different sense organs and possibly a very different natural and social environment from those of *homo sapiens*?

Should the first message we encounter be an intentional attempt on the part of an alien civilization to contact another intelligent species, it is reasonable to expect that such a message will be limited in content. Vakoch noted in an article in *SETI Explorer Magazine* that the few messages humans have sent into space have been rather limited, and a bit warped, in terms of their representation of our own civilization, showing largely the brighter sides of humanity and ignoring features such as war and poverty. Even if extraterrestrials are trying to represent themselves in an objective manner, it is very likely that any intentional message we receive will have subjective qualities and represent said alien civilization in a limited way that, in turn, will influence how we construct an understanding of their messages and, beyond that, their civilization and culture.

Regardless of the type of communication we receive, the fact is that humans are unlikely to receive a message and, then, sit back and simply take it at face value without speculating on the nature of those who sent it. Benedict, like armchair anthropologists before her, was a trained interpreter and theorist of culture and behavior with a PhD in anthropology from Columbia University. In short, she was an expert in collecting and analyzing data about cultures different from her own. But the conditions of her research on Japan and her lack of understanding of the Japanese language made it virtually impossible for her to gain a very accurate picture of the culture and people about which she wrote. Furthermore, her subjective interests in a particular theoretical framework influenced her management of the data she did obtain and led her to organize her understanding of Japan in a way that fit with her assumptions about how cultures work. This is natural, particularly when we are dealing with limited data. She did the best she could, given very difficult circumstances.

Furthermore, as was the case with Benedict's study of Japan, when it comes to contact with ETI this process will not be limited to a few scholars with training in understanding alien cultures. It will quickly become a forum for discussion among policy makers and be gradually released to the public, generating a broad, although perhaps temporary, public discourse most likely on both the nature of ETI and what it means for humans. Many of those initially discussing what it all means—such as astronomers—will be people who, in fact, lack even the type of training Benedict had, let alone the much more sophisticated and nuanced training that scholars such as anthropologists and sociologists receive today.[5] Indeed, as Shostak (2004, 2006) points out, should contact occur, knowledge of the event will become quickly evident to a wide audience, most likely well before SETI scientists are even certain that the signal is really from an extraterrestrial intelligence. Contact will become generally known about and reflected upon before anthropologists and other social scientists whose expertise is focused on the understanding of different cultures are able to analyze whatever content might exist in a signal. In short, the invention of an extraterrestrial alien culture will begin almost immediately upon the announcement that contact has been made—it will happen much more quickly than the invention of Japan did after Benedict's publication.

[5] I find this to be one of the more comical and disturbing aspects of quite a bit of science fiction literature on encounters with extraterrestrials. Sagan's *Contact* is a good example of this. The message is received and then a bunch of astronomers sit around trying to decode the message and figure out what it all means while trying to keep the government at bay. This is very odd, since astronomers have no more training in the collection, analysis, and interpretation of data on culture and social organization than do anthropologists or historians on the formation of stars. When I see the movie, in particular, I keep wondering where are the historians, anthropologists, sociologists, linguists, etc.? It's weird. Sagan does have a theologian in there for good measure, but it's pretty clear that Christian theologians at least are not exactly the most objective and scientific bunch on the planet.

In the case of encountering a message from an extraterrestrial intelligence, we are faced with the same problem that Benedict and earlier anthropologists working at a distance encountered. There will be limited data. In addition, we are also faced with the problem of a time lag—but not the lag of several months encountered by armchair anthropologists of the nineteenth century. Instead, we will encounter time lags of decades, centuries, or millennia between message and response. If we think about the course of the study of Japan, so heavily influenced by the work of Benedict even when access to new data has been readily available over only the past 70 years, it is fairly easy to imagine how long stretches with little or no data could lead to speculation and the creation of an imaginary related to the extraterrestrial civilization that is based largely upon our own theories about how culture and behavior work. Indeed, the vast majority of what we will "know" about ET if contact happens, is most likely to be our own invention based upon very limited data and, then, elaborated upon over the long periods of time that will exist between contact points.

5.3 New Imaginaries

In short, what the example of Japan anthropology following Benedict and other cases from the history of anthropology suggest is that even if social scientists, philosophers, and historians are involved, there is a good chance that humans will create an imaginary about ETI that is really a reflection of ourselves and our theories about how we organize ourselves culturally, politically, economically, and socially. That imaginary will, of course, also influence our own cultures and societies here on Earth, although I suspect it will happen in different ways depending upon which society one happens to be experiencing. The hopeful, and I think rather naïve, speculation among many who have thought about contact with ETI is that it will somehow bring us together as we think about our place in the universe and contemplate the meaning of encounter with an alien civilization. ET will be smarter than us, more advanced than us, and nicer than us. Maybe, if ET knows how to travel across the large distances of interstellar space, they will drop by and help us out of our "adolescent" stage so that we can become an "adult" civilization like ET. We are back to the Star Trek imaginary about contact, although every time I hear these types of ideas I think of the Twilight Zone episode *To Serve Man*.

I fully recognize that increasing unity is a possible outcome of humans becoming suddenly aware that they are not alone in the universe, although the reader probably notes the sarcasm in my tone. And, of course, many thinkers (both scholars and those writing in the area of science fiction) have pointed out that contact could just as easily throttle our society and cause chaos. Humans are pretty easily frightened. What I actually think will happen is an initially complex response that varies significantly from one human society to another and among different groups, with different interests, within various societies. For example, conservative Christians are likely to react quite differently from astronomers or anthropologists. Various groups will bring different value structures to the interpretive table and we will create multiple imaginaries about who and what ET is and who we are in relation to ET. These imaginaries will occupy the general public for a while and scholars perhaps for a long time. And then, given that the distances will be great and the time lag between "hi" and "how are ya" will be vast, we will return to our usual business of greed, poverty, hunger, and war. It will be a lot like landing on the moon.

But let's say I'm wrong—and I hope I am. What if we get really interested in ETI and that fascination starts significantly changing our society as we work through the import and meaning of contact? In that process, we need to be very careful not to let our hopeful ambitions and desires influence our ideas about who and what ET might be. I think it is very possible that ET will be from a fragmented world not unlike ours and will neither speak with a unified voice nor interpret what we might reply with a unified voice. In fact, if I had to bet on it, I'd go with this scenario. Why? Because the only empirical data on how "civilizations" evolve comes from right here on good old Earth—and nothing in those data support the idea that the long-term prognosis for a civilized society that is anything like us, and remember it may be necessary for aliens to be similar to earthlings if communication is going to work, is a unified world.

Greater technological innovation and abilities have done little to defragment our world. While there is a techno-bureaucratic elite, of which intellectuals tend to be a part and that is perhaps a bit more unified than much of our world, it is important for us to remember that most of the world isn't part of that elite. The people I watched on CNN this morning in Iraq who have suffered long as a result of the narrow vision and cynical power mongering of the American version of that techno-bureaucratic elite and who are now being pummeled by ISIS are but one good example of just how fragmented our world, with its capacity to intercept and send interstellar messages, actually is. And that gets us to the gist of our problem: *While we need to do all that we can to limit or avoid anthropocentrism in trying to figure out what a message from ETI might mean, we also have to recognize that the only data we can use at this juncture in history as a basis for trying to figure out what ET might be like is our own world.* Anything else is *entirely* speculative, because we are the only unambiguously intelligent creatures on this planet or, as far as we know right now, anywhere in the universe.[6]

To summarize, when anthropologists do research on other cultures, those places are not objectively studied, analyzed, and reported on. Rather, these cultures emerge or are created via an interplay of objective and subjective observations, assumptions, stereotypes, politics, misunderstandings and misinterpretations, and even personal egos that influence the ways in which the world of the alien other is represented both in scholarly literature and in public discourse (Ryang 2004: 29). Contact and

[6] In making this claim, I recognize that there are other intelligent creatures on Earth and that some of them may be quite intelligent. Dolphins are a good example. There has been some very interesting and important research looking at animal intelligence and communication as a model for communication with ETI (Herzing 2010; Doyle et al. 2011). But dolphins don't build radio telescopes and space ships, and it is likely that the type of extraterrestrial intelligence we might encounter will do those things and, thus, will be somewhat like us at least from a technological, if not from a social, perspective.

the likely very long-term spaces in interaction between extraterrestrials and humans (including SETI researchers, politicians, scholars outside of SETI, and the general public) will involve not simply the discover of an alien other, but a similar production of knowledge about that alien other that will generate a new imaginary about both extraterrestrial and human life as the discourse among numerous stakeholders unfolds surrounding the nature and motivations of ET. How far that imaginary goes to change our own world is an open question that is impossible to answer at this point in time. However, awareness of this, and the ability to reflexively think about our own role in constructing a representation of an alien culture, particularly where great distances and time delays are insurmountable with current technology, is of fundamental importance in thinking about how initial contact might change our own world.

Chapter 6
Religion, Science, Culture, and SETI

> *To talk of immaterial existences is to talk of nothings. To say that the human soul, angels, god, are immaterial, is to say they are nothings, or that there is no god, no angels, no soul. I cannot reason otherwise ... without plunging into the fathomless abyss of dreams and phantasms. I am satisfied, and sufficiently occupied with the things which are, without tormenting or troubling myself about those which may indeed be, but of which I have no evidence.*
>
> — Thomas Jefferson in a letter to John Adams, August 15, 1820

We have traveled quite an intellectual distance in this book, moving from the historical process by which modern humans became increasingly aware of and able to imagine the existence and even nature of extraterrestrial intelligence to a bit of speculation, based upon an example from the history of anthropology, about what might happen if we do, in fact, make contact. The central point of my argument has been that the defining feature of how our ability to contemplate ETI has grown and changed over time is tied to the ways in which culture, and the imaginaries associated with changing technological and ideational abilities and themes, is experienced among scientists and the general public. Science is embedded in cultural flows, and in the case of SETI those flows have been closely tied to ideas of technological and cultural progress that are a product of the interplay of Western religion and science.

In this chapter, I turn more directly to the question of the relationship between SETI, religion, and culture. The quotation above from Jefferson is interesting to me because, although he was writing about religion, one can modify his comment slightly and add to his list ETI when he talks about the immaterial beings that are nothings without empirical evidence. As much as I want to believe that there are other civilizations out there, until we make contact and have empirical evidence of their existence, we are stuck with the reality that they are nothing at all or if

J. Traphagan, *Extraterrestrial Intelligence and Human Imagination*, Space and Society, DOI 10.1007/978-3-319-10551-2_6 anything, nothing more than a hope and desire on the part of humans. They are phantasms, or products of the imaginary that we have created to represent a hoped-for other who shares the vastness of the universe with us. But the evidence does not support their existence, nor after the limited time we have been looking does it support their absence. We simply don't know. And it's in that unknowing and our response to it that I see connections between SETI and religion as intertwined cultural phenomena.[1]

Before expanding on this point, however, let's use a few paragraphs to restate the main goals I set out for this book in Chap. 1. My primary ambition has been to explicate some of the underlying assumptions that have influenced both our capacity to think about the existence and nature of ETI and the ways in which SETI researchers conceptualize the purpose of their endeavor and represent ideas about the potential influence contact might have on human societies. A couple of themes can be drawn from the discussion.

First, scientists in general and SETI scientists in particular, do not do research in a vacuum of objectivity shielded from the subjectivities of culture and emotion. Just like the fact that the general public's ability to grasp the idea of ETI has been shaped by changes in technology and knowledge of the universe, so has the imaginary of astronomers and other scientists interested in the possibility of extraterrestrial life. While all science is embedded in cultural context, I think this reality is somewhat more robustly influential in the area of SETI, because SETI scientists are involved in a topic that is of wide interest to the general public and is routinely presented in pop culture in fictional and non-fictional products of our postmodern imaginary. SETI scientists also ask questions that significantly overlap with those of the religious ilk, making them both an object of fascination and a target for those interested in theology. This demands on the part of SETI scientists a regular awareness of and response to the questions, ideas, attacks, etc. of those interested in what SETI scientists are doing and what contact with ETI might mean for humanity.

[1] I find it fascinating how easily humans can take a lack of information and turn it into evidence for something. For example, in a recent, and interesting, article on the issue of risk assessment and SETI, Neal (2014: 67), drawing upon the work of Petigura et al. (2013) states that, "Scientific evidence is thus increasingly supportive of there being life of some kind in the universe—and this includes life on other planets in our own galaxy, the Milky Way. With every new finding of rocky exoplanets in the "Goldilocks zones" of nearby stars, the null hypothesis—that there is no life in the universe but on our own planet—looks more unlikely and untenable." Unfortunately, this position is the product of a logical mistake. There remains no more evidence for the existence of extraterrestrial life today than there was 100 years ago or 1,000 years ago. Therefore, we cannot draw any kind of conclusion about the likelihood of extraterrestrial life based upon the data we have to date. What we do know is that rocky planets like Earth appear to be very abundant, giving us *hope* that some of those planets harbor life and even intelligent life. But there is no evidence for anything other than the existence of millions of small rocky planets with similarities to Earth in our galaxy. Should we find, for example, indications of fossilized microorganisms on Mars, then that's a potential game changer, because it suggests life may have formed twice in our own solar system, indicating the possibility of life forming often. And even that conclusion is open to question if one theory, that life here came with ejecta from Mars and populated Earth, is correct.

Indeed, this has been a major point of my writing thus far—the imaginary in which SETI resides is one that is deeply intertwined with both the imaginaries of Christian theology (at least, and maybe others, as well) and the imaginaries of science fiction. I am neither arguing that SETI is a new religion nor that it is nothing more than science fiction; but the intellectual and cultural domain it inhabits is situated close to and often congruent with these areas of human thought and creativity. We shouldn't neglect this fact when thinking about how SETI scientists are influenced by their cultural milieu, nor when thinking about the role SETI plays in public discourses related to the nature of humans, space exploration, and religion or the potential impact contact with ETI might have on both scientists and the general public. From a social perspective, SETI is a node in the middle of that discourse and a broad imaginary of contemporary life that centers hope for the future on the technological innovations of the present.

A second theme is related to the ways in which SETI scientists have tended to imagine the nature of both human and potential other civilizations. This is an area where I think we can engage in a bit of constructive criticism, because it is clear that many in the SETI community have developed their thinking about the nature of extraterrestrial civilizations upon a flawed understanding of the history of human civilizations, and the nature of human culture, that works from assumptions about cultural evolution which are simply not accurate. While we can clearly see technological innovations becoming increasingly complex over time (even while there are periods where they regress, as well) and it is clear that as human populations increase, the level of complexity of their political organization tends to become more layered and hierarchical, there is no empirical basis for linking technological progress or complex forms of social organization with moral or social progress. It is very difficult to find any good reason to think that human individuals or human societies are in any way a moral or social improvement over our ancestors. To be sure, we are *different* in many ways from past humans. The size and complexity of our current forms of social organization and the enormous influence our technology has in managing how human bureaucracies operate and many people live is different from the past, but at both individual and collective levels we continue to be a mix of both nice and not-so-nice qualities that haven't changed much over long periods of time. We still are capable of deep kindness and powerful hatred, of generosity and greed. And we have shown no significant trend toward a reduction of institutional and other forms of violence, even after fighting two horrific world wars. I see little evidence that we've made a great deal of moral progress throughout our history, but we have found a variety of new ways to both inflict harm on others as well as to cure those in pain and suffering.

The interesting thing here is that biological evolution does not assume progress, either. It's a theory of change. There may be increased complexity of some organisms, but simple single-celled organisms continue to be important and, in fact, dominant forms of life from a population perspective. Biological evolution does not change with directionality or a *telos*, it is just a process of change. There's no target. What needs to be developed much more robustly in SETI research is an awareness that the same is the case for cultural evolution. Cultures evolve in the sense that they change, but they aren't going anywhere. It's just a non-directional process of change with neither progress nor regress, even while we can observe characteristics like increased complexity of social organization as populations grow. That increased complexity is a byproduct of larger populations, not evidence of a direction toward which cultures or civilizations inevitably move.

In many ways, the tendency to assume that cultural evolution is directional is a product of the teleological tendencies of Western philosophical and theological thought in which causality seems to be moving the world toward a particular, and often predetermined, effect. In Christian theology, this is expressed through the idea of the second coming of Christ and the end of this world order to be replaced by a utopic follower of the apocalypse. This linear worldview has been profoundly influential in Western conceptualizations about the order and historical nature of the universe, and has contributed to feelings among many that the order, structure, and seeming directionality to change is evidence for the existence of some sort of superior intelligence such as the divine watchmaker of deism or the Christian god that intentionally created the universe and the laws by which it operates. This mindset is very well expressed in the work of Thomas Acquinas, who argues that:

> We see that things which lack knowledge, such as natural bodies, act for an end, and this is evident from their acting always, or nearly always, in the same way, so as to obtain the best result. Hence it is plain that they achieve their end, not fortuitously, but designedly. Now whatever lacks knowledge cannot move towards an end, unless it be directed by some being endowed with knowledge and intelligence; as the arrow is directed by the archer. Therefore some intelligent being exists by whom all natural things are directed to their end; and this being we call God (Aquinas, *Summa Theologica*, Article 3, Question 2, cited in Elders 1990: 120).

The type of theology we see in the ideas of Acquinas, as well as in the divine watchmaker analogy of the universe made famous by theologian William Paley in his work of Christian apologetics, *Natural Theology* (1802), and which holds that the complexity of living organisms is evidence for the existence of a divine creator, has within it assumptions that lurk in the ways in which many SETI scientists imagine ETI. It's not that SETI scientists are arguing in favor of the existence of a divine entity of any sort; rather, it is that they tend to work from tacit assumptions about progress and directionality in the development of intelligence and culture that are products of the same worldview that has shaped Western theology and philosophy and that are grounded in the musings of thinkers like Acquinas, Paley, and many others who see the world as having a conscious creator, a purpose, and an identifiable direction to the flow and terminus of history. This way of seeing the world is not supported by the science of evolutionary biology, which neither presupposes intelligent design nor any directionality beyond the possible changes that can be successful within the limits of what has already evolved (Dawkins 1986). Nor is it supported by the understanding of cultural change that has developed among anthropologists over the last century of ethnographic research on human behavior and social organization.

Again, I'm not trying to suggest that SETI scientists are closet theologians. Rather, I am arguing that their way of thinking about the nature of ETI is influenced by the imaginary of Western cultural and religious ideas and values, with their strong emphasis on linear social change, particularly evident in the social Darwinism associated with cultural evolution, despite the fact that empirical evidence on Earth shows that social change is non-directional, just like biological change.

The same basic features of human behavior and social organization continually appear throughout our history, even if they are balanced in different ways. I think modern democratic societies are better to live in than authoritarian societies, but we aren't the first to be democratic and those same democracies are capable of considerable nastiness—we still fight wars and have poverty. I was born in 1961 and there is no extended period of time since my birth that the US has not been engaged in some sort of military operation, many of which have been long-term wars (Vietnam, Iraq I and Iraq II, Afghanistan, and so many smaller engagements) in which the moral compass and motivations of the US government have been open to considerable debate.

The point? There are many *different* types of social and political organization for humans and most of them co-exit on our planet at the same time. What we don't seem to see in human history is one type of social and political organization supplanting all of the others—and, no, it is not at all clear that such a change is happening now with globalization, either (see Traphagan and Brown 2002). Nor can we argue that one form of social organization is the product of a linear progression toward some higher level of social, moral, or cultural way of living.

We do see large, state-level societies that overrun tribal societies, for example, but then we see other forms of tribal-type social organization emerge within the larger state-level societies—think about parts of the Middle East or group formation in corporate offices. And in some cases, such as the San bushmen in Africa, a hunter-gatherer lifestyle was maintained until fairly recently and with that what seems to have been a relatively egalitarian relationship between men and women. But Becker's research has shown that "as a result of specific social and historical circumstances, distinct and hierarchically organised [sic] perceptions of 'men' and 'women' have begun to establish themselves to varying degrees among southern African San communities" (2003: 8). Becker goes on to argue that violence between men and women among the San has been exacerbated as gender hierarchies and a variety of historical and social characteristics influenced by the larger context they inhabit that has influenced San social organization and gender relations as they have been forced to live within the framework of a state society.

What this means for SETI is that we must be very cautious about not infusing culturally shaped notions of progress *qua* cultural evolution into our thinking about either our own or any potential extraterrestrial society. If we take Earth as our only data point, we don't have any evidence to suggest that ET will have somehow culturally evolved to a higher level of social organization that allows only for altruism and general niceness towards less developed or "adolescent" species like humans. We *do* have evidence that contradicts this assumption in the endless stream of human societies that have overrun and decimated other societies that controlled less sophisticated types of technological development than their intentional or unintentional conquerors who did not appear to harbor corresponding improved abilities when it comes to moral action.

This is a locus, I think, where the imaginary of science fiction has significantly influenced the ways some SETI scientists and many in the general public interested in SETI think about the nature of extraterrestrial intelligence. As noted in the previous chapter, this Star Trek imaginary is one in which there is a tacit assumption that as a species develops increased technological capabilities, social and political forms develop into new, *and better*, types that gradually supplant unpleasant behaviors of the past. Star Trek, of course, is a product of Western society and reflects *American* optimism about the potential of humans to "overcome" primitive behaviors (remember, however, that the egalitarian social organization often expressed in American political ideology is actually far more common in so-called primitive societies than it is in modern, technologically advanced societies like the US)—Captain Kirk always seems to be showing us that we can get beyond our baser instincts and that there is hope for us in the long-run. And that American optimism is itself closely tied to Christian conceptualizations of the potential for human release from sin and ultimate entrance into a future utopia devoid of evil.

I like optimistic views of our future and I hope that's what will happen. But at the moment all we have is an abiding faith either in some kind of spiritual enlightenment (for many Christians and others) or technological innovation (for humanists and various religious people) to solve our social and political problems. However, the empirical evidence from human history does not support optimism about the validity of that faith at this point in time. The SETI endeavor is grounded in that faith in some ways, because it works from the assumption that if contact is made, bad things won't happen to us because ET is going to be more advanced than we are technologically (likely true) and, therefore, it will be morally more advanced than we are (no idea if this is true or even possible, but no evidence-based reason exists to think it is). Of course, there are detractors who disagree (Stephen Hawking is one) that ET will be nice, but it seems unlikely that anyone would seriously engage in seeking out contact with extraterrestrials they truly believe would be monsters bent upon beating the daylights out of adolescent earthlings.

If you are not yet convinced that the idea of cultural evolution as directional and progressing is in fact a cultural product, keep in mind that this is not a necessary way of seeing the world. In Buddhism, for example, there is no god, no arrow of time toward which history marches, and no sense that one form of human social organization is better than other forms. From the Buddhist perspective, the idea that the world is going somewhere is an illusion generated by our desire to cling to a sense of permanence in our relationships to others, the nature of our surroundings, and to our own sense of self (see Kopf 2001). Buddhist philosophy argues that everything is always changing, but that there is no endpoint toward which that change is directed. It's just change. Sound familiar? This is obviously a very truncated and simple expression of Buddhist philosophy—it would take many more chapters to do justice to this issue. My point is simply that in two different cultural and historical contexts we find different ways of thinking about time, change, progress, and the fundamental nature of the universe and that SETI is influenced, significantly, by the Western/American context in which it happened to develop and become a significant part of the imaginary of life here and on other planets.

The second broad area we have explored in this book is the question of what the search for extraterrestrial intelligence tells us about ourselves in terms of the types of values we express related to what constitutes intelligence and civilization. In large part, SETI is guided by an imaginary that is deeply situated in our own assumptions about what constitutes intelligence and an anthropomorphic representation of the likely nature of ETI. Even when we consider mathematics as a potential means to communicate, we are faced with the fact that it is a symbolic system that is grounded in our own being—it is related to both our biological and cognitive nature and, thus, is limited by human cognition.

This presents a very difficult problem for SETI, because it raises at least the possibility that our world and that of ETI might be incommensurate. Of course, if that is the case, we probably won't bump into each other anyway (at least via radio signals), so it probably doesn't matter. But it is important to think about this limitation, because it helps us to recognize the extent to which we are inclined to tacitly anthropomorphize both the nature of ETI and the means by which we conduct the search. This becomes a limiting factor that may inhibit the search, but is also a challenge to avoid because it is extremely difficult to imagine communication with something dramatically different from ourselves.

Some scholars of SETI have recognized this problem and speculated that ETI, given the likelihood of it being a long L civilization (meaning very old) would have evolved into a post-biological state, being perhaps entirely a form of "artificial" intelligence or an evolutionary step beyond biology-based intelligence (Dick 2008). This makes a great deal of sense and certainly provides one path to take us away from imaginaries related to the nature of extraterrestrial civilizations that are dependent upon the linear progress model of cultural change. But it also raises a host of new questions, the most intriguing being, in my opinion, whether or not a post-biological being would even recognize us as intelligent. One might imagine such a being as an integrated AI in which there is no concept of individuality and, thus, the idea of our form of life as a collective of separate bodies might not sufficiently link in to the post-biological being's idea of intelligence or culture for it to recognize us as having either. This is certainly an interesting avenue for future thought.

Finally, the topic I have not raised to this point centers on the question of the extent to which occasional claims that SETI is itself a kind of faith-based religion represent a reasonable, or unreasonable, critique of the field. This moves us into an area of inquiry that has been quite fertile for the past 50–70 years or so and has included scholars from a wide range of disciplines, including physics, theology, philosophy, and the social sciences, although in general social scientists have not been particularly interested in this topic. This may be a product of the fact that scholars in some areas of the social sciences have been preoccupied with debating whether or not what they do even should be considered scientific or whether, as sciences, they should be considered on equal scientific footing with the natural sciences (cf. Kincaid 1990). Anthropologists, in particular, have been concerned with debating the nature of their discipline around the question of whether or not the word science should even apply to what they do for a living. In 2010, the American Anthropological Association (much to the consternation of many of us who are members) decided to remove the word "science" from a statement describing the discipline. The decision was related to a long-standing debate among those of us, such as many archaeologists, physical anthropologists and some cultural anthropologists like me, who are inclined to see anthropology as a science-based discipline and those who see it as a means to advocate on behalf of human rights for native peoples (these scholars have generally been situated in cultural anthropology and their research tends to focus on issues of race, ethnicity, and gender). I am not interested in engaging the rather long, and sometimes quite intense, debate among anthropologists about this issue; suffice it to say that I see myself clearly as a scientist and am not partial to the strain of activist anthropology that has moved the discipline away from scientific inquiry, even while I am often sympathetic to the causes that some activist anthropologists promote. Despite the presence of this issue in the social sciences, there remains a long-standing intellectual discourse exploring both historical and epistemological connections between scientific reason and religious thought more generally throughout parts of the academy.

6.1 Science and Religion

There have been quite a few scholars who have explored substantive similarities between religion and science including philosophers of science (Margenau 1984), religious studies scholars (Barbour 1997), theologians (Pannenberg 1993), working scientists (Schrödinger 1983 and also Margenau 1984), and even a few scientist/theologians (Polkinghorne 2007). Others have tried to equate aspects of religion and science, including some pretty wacky stuff like physicist Fritjof Capra's best-seller *The Tao of Physics: An Exploration of the Parallels Between Modern Physics and Eastern Mysticism* (1975), which not only ushered in a wide range of largely silly "Tao of" books, but showed an impressively superficial understanding of Eastern mysticism while also being criticized by many in the physics community (although some also praised the book), often on the grounds that Capra simplistically equated similarities of language without understanding the full meanings implied either by physics or mysticism. No matter; he sold a lot of copies and many people bought into the basic idea of his book.

Theologian/philosophy Ian Barbour, who held undergraduate and graduate degrees in physics, devoted his career to trying to sort out the historical and conceptual intersections between religion and science. In his last book, Barbour (2013a) notes that science and religion got along O.K. during the seventeenth century, but over time there have been a variety of frameworks that have represented the relationship in four ways, including conflict, independence, dialogue, and integration, with the conflict theme becoming characteristic in the second half of the twentieth century.

Although, his work is characterized as being about science and religion, Barbour, for the most part, was less interested in the relationship between science and religion than he was in the relationship between science and Christianity. In fact, this is a widespread feature of the science/religion debate—it is usually really a science/Christianity debate. As noted above, there is little basis for conflict between religions like Buddhism or Taoism and science, since neither of these have absurd mythologies about the world being created in 6 days or the Earth being less than 10,000 years old. Nor do Buddhism and Taoism have a focus on the idea of creation or the origin of the universe. For Buddhism, as noted above, the physical universe is an illusion—in essence, the question of where the universe came from is irrelevant because it is "created" by human misunderstanding that our suffering is generated from our desire to cling to the permanence of the world which, in truth, is constantly changing. Furthermore, there is no basis for conflict between Buddhism or Taoism and scientific frameworks such as evolutionary theory or astrophysics, because these religions don't concern themselves with the idea of a particular individual who happened to decide to create the universe for fun one day. It's no surprise that it is quite uncommon to find people from countries influenced by Buddhism and Taosim who question the verity of evolutionary theory.

6.2 Is SETI a Religion?

This brings us to a question that has been raised by a variety of individuals, many of whom want to claim more generally that science is based on faith and, thus, is no different from Abrahamic religions. Science is a method for understanding the world, it is not based on faith due to the simple fact that, unlike faith-based approaches to understanding, it is inherently open to challenge and revision. As I argued in Chap. 1, scientific knowledge is inherently contingent. Nonetheless, quite a few people mostly from conservative Christian contexts have tried to argue that science in general is faith-based and that SETI in particular is actually a kind of religion. Tom Bethell, a conservative journalist who writes on religion, economics, science, etc. is one such person, but others with a less religiously-oriented approach, such as Jurassic Park author Michael Crighton, have also argued that SETI should be viewed as a religion. In a 2011 article in the *American Spectator*, Bethell writes:

> What [SETI] scientists are looking for, of course, is extra-terrestrial life, not rocks orbiting stars. The late novelist Michael Crichton gave an entertaining lecture at Caltech in 2003 saying that the search for extraterrestrial intelligence is a religion. And in a way it is. Carl Sagan, one of its leading promoters, "believed in superior beings in space, creatures so intelligent, so powerful, as to resemble gods." He affirmed that a new civilization is formed just in our galaxy every 10 years. "There are a million technical civilizations in the [Milky Way] galaxy," he believed. That's religion.[2]

No, it's not. The belief that there may be beings in the universe superior to humans does not necessarily imply religious belief; in fact, it has nothing to do with religious belief *per se*. This is akin to stating that a belief that there is a planet orbiting the sun beyond the orbit of Pluto is a religious belief and stems from the conflation

[2] http://spectator.org/articles/36734/extraterrestrial-intelligence-and-search-god, accessed 11/22/2013, 10:40 am.

of belief as it relates to hypothesis formation to belief as it relates to faith. Again, the first is open to challenge and change, the second isn't.

Unfortunately, arguments—and I use that term very lightly—like Bethell's that try to represent SETI as a religion normally lack a very clear definition of religion and assume that Abrahamic religions are the only types of religions that exist. This is perhaps not surprising, since even scholars of religion have a difficult time defining the object of their study. However, the position taken by Bethell shows a faulty understanding of both religion and science and conflates the *belief* that something may be possible or is even highly likely with *faith* and *conviction* that it is unquestionably real. I can't imagine any scientist rejecting the possibility that humans are alone, precisely because we have no inconclusive evidence that we are not alone.

Regardless of what Sagan may have believed about the abundance of extraterrestrial life, he was, in fact, quite interested in the relationship between the scientific and the religious search for our place in the universe. His novel, *Contact*, directly explores this issue:

> You see, the religious people—most of them—really think this planet is an experiment. That's what their beliefs come down to. Some god or other is always fixing and poking, messing around with tradesmen's wives, giving tablets on mountains, commanding you to mutilate your children, telling people what words they can say and what words they can't say, making people feel guilty about enjoying themselves, and like that. Why can't the gods leave well enough alone? All this intervention speaks of incompetence. If God didn't want Lot's wife to look back, why didn't he make her obedient, so she'd do what her husband told her? Or if he hadn't made Lot such a shithead, maybe she would've listened to him more. If God is omnipotent and omniscient, why didn't he start the universe out in the first place so it would come out the way he wants? Why's he constantly repairing and complaining? No, there's one thing the Bible makes clear: The biblical God is a sloppy manufacturer. He's not good at design, he's not good at execution. He'd be out of business if there was any competition (Sagan 1985: 285).

Clearly, Sagan doesn't have much use for the god of the Abrahamic religions, I'm inclined to agree, but that is a different discussion.

There are quite a few quotations attributed to Sagan that focus on the issue of religion (http://atheism.about.com/library/quotes/bl_q_CSagan.htm), and most of these conceptualize religion in terms of the Abrahamic tradition in which the god in question is an all-powerful, creator-god that seems to have invented the universe and consistently tinkers with it for no apparent reason other that getting kicks out of messing with its play things.

The important point to keep in mind here is not the problematic nature of the Abrahamic god, but that "religion" for most of those involved in SETI research and those who critically identify SETI as a religion seems to mean religion in the Abrahamic sense of the word. In other words, it is Western-style religion, which depends upon the idea of blind faith and emphasizes dogmatic sets of rules that govern the scope of that faith and the behavior of those who adhere to it. Of course, this is only one model for religious behavior that we have, but it is the model of religion that is typically used when attempting to label SETI as a religion. So the question of whether or not SETI is a religion, addressed both by those who support the idea and who reject it, is really one of whether or not SETI is a faith-based religion along the lines of Christianity, Judaism, and Islam.

6.3 SETI, Religion, Imagination

SETI does appear to have some things in common with the Western approach to religion. First, the object of attention (god or ETI) is a thing that is entirely a product of our imagination in both cases. There is no evidence for the existence of a god, nor is there any evidence for the existence of ETI. But that's where the similarities on this point end. Western religious types cling to the faith that their hoped-for deity must necessarily exist, in spite of the lack of evidence to support that conclusion. Some will argue that the universe itself is evidence for the existence of a creator—how would it get here otherwise? But if one takes that approach, then one has to ask who or what created the creator? And who or what created the creator that created the creator? And so on—it's turtles all the way down. Arguing that there must be a creator due to our very existence or due to the seemingly designed nature of the universe doesn't get us any closer to an answer to anything, because we can always just push the question backward.

In this sense, SETI is nothing like the Abrahamic religions, because it does not require a positive answer to the question of the existence of the alien other (whether that alien be ETI or a god). The answer can be "no." That said, many in the SETI community operate *as though* theirs is a religious quest, in that they hold to the conviction, as did Sagan, that there just ought to be someone else out there. It seems strange that Earth would be the only place that intelligent life would arise in a galaxy as vast as ours, let alone an infinite universe. But again, religious faith of the Western variety demands a positive answer. One cannot have faith and admit of the possibility that his or her faith is completely misguided. In this sense, SETI cannot be defined as a Western-style religion, because it carries within it the basic scientific approach of falsification. While there is hope in SETI—many SETI scientists, myself included, hope that there are other intelligent beings out there—there is no faith that they must, by definition, be out there. If the answer turns out to be that we are alone, then that's the answer. The problem lies in the fact that because our chances of finding ETI are slim and the amount of time needed to listen for ETI in order to get a good answer is necessarily calculated in decades and even potentially centuries, SETI has a tendency to take on the appearance of a faith-based religion focused upon an intangible nothing of the sort that Jefferson eloquently derides.

There is a further problem that contributes to this religification of SETI among some critics and that is largely generated out of the manner by which many scientists talk about their work. In his discussion of the conflict/relationship between religion and science Barbour (2013b, Kindle Locations, 302–310) argues that when scientists attempt to discuss religious beliefs, they tend to run into problems because they do not adequately distinguish philosophical from scientific questions.

> Scientists, in their popular writings, tend to invoke the authority of science for ideas that are not really part of science itself. Theism and materialism are alternative belief systems, each claiming to encompass all reality. In their epistemology, these authors assume that the scientific method is the only reliable source of knowledge—an assumption sometimes referred to by its critics as scientism. If science is the only acceptable form of understanding, explanation in terms of evolutionary history, biochemical mechanisms, or other scientific theories excludes all other forms of explanation.

For Barbour, belief in the Christian god, at least, is "commitment to a way of life" that occurs in response to religious experiences. Religion, as such, offers a framework for meaning through which historical events can be contextualized and, therefore, it is really quite different from science, which generally does not attempt to provide that type of framework, even while many scientists themselves engage in the sorts of philosophical musings associated with religion.

A good example of this can be found when SETI scientists discuss the potentially world-changing event that contact with ETI might represent for humans. This, of course, is a philosophical question rather than an empirical question—it exists within the realm of speculation about what might happen in terms of the *construction of meaning* about our species and its place in the universe. And as a question of meaning, it really cannot be addressed with the methods of natural science (this is also why the Rio Scale doesn't measure anything). I don't think it can really be addressed with theological methods either, since Western theology is based upon assumptions about the universe that don't make a great deal of sense, but it remains that questions of meaning do not lend themselves well to natural scientific approaches to understanding the universe. The best place to contemplate these types of questions is within the social sciences, which have a well-developed set of methodologies to explore how people construct meaning. That said, I've made a subtle shift in the last sentence, because I've moved from what it all means to a question of how humans think about what it all means. From my perspective as a social scientist, we aren't going to get answers to the theological and philosophical types of questions because the answers are cultural and contingent. The contingent nature of meaning raises the point that social science is also situated in opposition to the philosophical and theological areas of inquiry.

However, I think Barbour is right about the problems many scientists face when venturing into the realms of theology and philosophy. When it comes to SETI, it is not a religion, but SETI scientists, perhaps due to the rather grand nature of their endeavor and the frustrating lack of evidence that characterizes their search, easily drift into discussions that have philosophical or even theological implications—without any data to discuss, it is easy to spend a great deal of time contemplating the philosophical/social implications of one's work.

This issue is related to the distinct types of questions scientists and theologians ask. Scientific questions focus on causes that lead to a particular condition. Religious questions are of "a much more burning, personal sort…because we are asking about the ultimate security, the meaning and destiny of our own existence" (Gilkey 1965: 19). The problem here is that SETI scientists work along the clearly defined framework of science in terms of *how* they do their work. But when they talk about *why* they do their work they often drift into this other realm of burning, personal questions about meaning and destiny that can be found in comments about how SETI can help humans understand our place in the universe or the significance of initial contact for humanity. In essence, SETI in its scientific form won't do that. It will tell us whether or not we are alone. It will be left to philosophers and theologians, most likely, to determine what it all means for us should contact be made.

I see no reason why SETI scientists shouldn't ask these philosophical and theological questions, but they need to recognize that as soon as they move into the realm of meaning, they are drifting into the domains of philosophical and theological thought and thus open themselves up to the charge that SETI is at some level a religious quest. Along the same line, when SETI scientists speculate on the social impact of contact for humans, they are drifting into the area of social science, which has well-defined methodologies and practices associated with thinking about human social organization and how people in different cultures construct meaning.

The potential desire to explore issues of the implications of scientific inquiry, of course, are evident in any science that deals with big questions. Physicists concerned with the origins of the universe are asking about how we got here and, thus, encroaching upon some religious people who believe that they already have the answer to that question. When evolutionary biologists explore how humans and other organisms came into being, they are also asking questions that Western religious types, at least, think they have already answered.

6.4 SETI and the Western Worldview

What SETI and Western religions certainly do share is that they are both products of a particular set of assumptions about how we know what we know in the world, even while these assumptions are different. SETI is grounded in the methodologies of empiricism that are, themselves, philosophical stances of an epistemological kind, albeit stances that are in direct conflict with the intuition-based epistemologies of religious traditions. The notion that scientific method arrives at accurate or at least reasonable understandings of the world *is* an epistemological position about the nature of how we know things in the world, even if we grant that it is the best epistemological position based upon its evident success in explaining and predicting the operations of the universe. Furthermore, one can claim, as I am doing here, that SETI and other sciences are products of cultural context and, thus, are shaped by cultural and historical processes and ideas that change over time.

Although SETI is clearly not a religion, the central argument to this book is that it *is* a cultural product and that it needs to be understood not simply as a scientific endeavor, but as a particular kind of scientific endeavor situated within an imaginary about humans and others that has developed since the Enlightenment and that is deeply shaped by a variety of Western values drawn from regions of thought such as Christian theology and social Darwinism. What this process has generated is a frame of scientific inquiry that tends toward anthropomorphism both in the way we think about the nature of ETI and the way we think about the nature of extraterrestrial civilization and is deeply influenced by Western assumptions about the nature of change as being progressive, directional, and leading to some sort of ultimate end. The imaginary associated with extraterrestrial intelligence that has emerged over the past two centuries or so is one that is powerfully built upon the ways in which Westerners and Americans think about ourselves both as individual creatures and as a "civilization." In some ways, the entire philosophical arm of SETI is a deeply anthropomorphic endeavor, because so much of it is built, either consciously or tacitly, around the only empirical evidence for life that we know of—our own world. And most interestingly, a significant element of its anthropomorphism is grounded in ideas about change that are products of the Western theological and philosophical traditions.

It is difficult to avoid this problem because all ways of looking at the world, scientific or non-scientific, are products of historical processes and specific cultural contexts. But for SETI to continue to develop it is important that natural scientists, social scientists, and those in the humanities interested in this topic consciously work to pull the imaginary in directions that move away from a very human-centered and ethnocentric (meaning centered on Euro-American notions of cultural evolution and Christianity) construction of the potential extraterrestrial other.

Finally, as I bring this to a close and as I noted early in the book, despite the fact that for intellectuals and in the broader context of human history, contact with ETI is certainly a highly significant moment, I remain unconvinced that contact with ETI would actually represent a major transitional point for humans in terms of the practical realities of living on Earth. I'm not sure it will mean much to most of the human population who struggle for survival. It will be interesting for a while, but it may well be that quickly people will return to the business of survival, war, poverty, politics, etc. I'm not arguing against the importance of SETI, but I am questioning its likely impact in the long-run. In many respects, SETI is more about us than it is about extraterrestrials. It tells us a great deal about what scientists, in particular, think about what is happening here on Earth and what it means to be intelligent. And our ongoing discussions about the nature of ETI, the meaning of contact, and the difficulties of communication are windows into an imaginary that situates us both in relation to each other and to the universe. The gaze of the SETI imaginary is focused as much on our own world as it is on the stars.

References

Abu-Lughod, L.: Writing against culture. In: Fox, R. (ed.) Recapturing Anthropology: Working in the Present. School of American Research Press, Santa Fe, NM (1991)
Ackerknecht, E.H.: Review of the Chrysanthemum and the sword. Q. Rev. Biol. **22**(3), 246 (1947)
Almár, I., Tarter, J.: The discovery of ETI as a high-consequence, low-probability event. Acta Astronaut. **68**(3–4), 358–361 (2009)
Appadurai, A.: Modernity at Large: Cultural Dimensions of Globalization. University of Minnesota Press, Minneapolis, MN (1996)
Barbour, I.: When Science Meets Religion. HarperCollins, New York (1997)
Barbour, I.: Religion and Science. HarperCollins, New York (2013a)
Barbour, I.G.: When Science Meets Religion: Enemies, Strangers, or Partners? HarperCollins, New York (2013b). Kindle Edition
Becker, H.: The least sexist society? Perspectives on gender, change and violence among southern African San. J. South. Afr. Stud. **29**(1), 5–23 (2003)
Benedict, R.: Patterns of Culture. Houghton Mifflin, Boston, MA (1934)
Benedict, R.: The Chrysanthemum and the Sword. Houghton Mifflin, Boston, MA (1946)
Bernard, H.R.: Research Methods in Anthropology. AltaMira, New York (2011)
Boas, F.: The Mind of Primitive Man. MacMillan, New York (1921)
Bourdieu, P.: Outline of a Theory of Practice. Richard Nice, trans. Cambridge University Press, New York (1977)
Bourdieu, P.: The Logic of Practice. Richard Nice, trans. Stanford University Press, Stanford, CA (1990)
Brown, M.: Radio mars: the transformation of Marconi's popular image, 1919-1922. In: Winn, J.E., Brinson, S.L. (eds.) Transmitting the Past: Historical and Cultural Perspectives on Broadcasting. University of Alabama Press, Tuscaloosa, AL (2005)
Capra, F.: The Tao of Physics: An Exploration of the Parallels Between Modern Physics and Eastern Mysticism. Shambhala, Berkeley, CA (1975)
Cartwright, J.: Evolution and Human Behavior: Darwinian Perspectives on Human Nature. MIT Press, Cambridge, MA (2000)
Chandy, L., Kharas, H.: The contradictions in global poverty numbers. Brookings. http://www.brookings.edu/research/opinions/2012/03/06-contradictions-poverty-numbers-kharas-chandy (2012). Accessed 9:15 am, 18 Jan 2013
Clyde, P.H.: Review of the Chrysanthemum and the sword. Am. Pol. Sci. Rev. **41**(3), 585–586 (1947)
Darwin, C.: On the Origin of Species. John Murray, London (1859)
Dawkins, R.: The Blind Watchmaker. W. W. Norton, New York (1986)

J. Traphagan, *Extraterrestrial Intelligence and Human Imagination*,
Space and Society, DOI 10.1007/978-3-319-10551-2

Dawkins, R.: The God Delusion. Mariner, London (2006)
De la Fuente Marcos, R., de la Fuente Marcos, C.: On the recent star formation history of the Milky Way disk. New Astron. **9**, 475–502 (2004)
DeVito, C.: On the universality of human mathematics. In: Vakoch, D.A. (ed.) Communication with Extraterrestrial Intelligence, pp. 439–448. State University of New York Press, Albany, NY (2011)
DeWitt, R.: Worldviews: An Introduction to the History and Philosophy of Science. Wiley, Malden, MA (2011)
Dick, S.J.: Life on Other Worlds: The 20th-Century Extraterrestrial Life Debate. Cambridge University Press, Cambridge (1998)
Dick, S.J.: The postbiological universe. Acta Astronaut. **62**(8–9), 499–504 (2008)
Dickson, P.: Sputnik: The Shock of the Century. Bloomsbury, New York (2001)
Doherty, T.P.: Cold War, Cool Medium: Television, McCarthyism, and American Culture. Columbia University Press, New York (2013)
Doyle, L.R., McCowan, B., Johnston, S., Hanser, S.F.: Information theory, animal communication, and the search for extraterrestrial intelligence. Acta Astronaut. **68**(3–4), 406–417 (2011)
Drake, F.: Encoding altruism. SETI Institute website. http://www.seti.org/seti/projects/imc/encoding/altruism.php (2008). Accessed 17 Mar 2008 (10:00 am)
Durant, W., Durant, A.: The Story of Civilization, Volume VIII: The Age of Louis the XIV. Simon & Schuster, New York (1963)
Durkheim, E.: The Elementary Forms of the Religious Life. Dover, Mineola, NY (2008 [1915])
Elders, L.J.: The Philosophical Theology of St. Thomas Acquinas. E. J. Brill, Leiden (1990)
Eldrege, N., Gould, S.J.: Punctuated equilibria: an alternative to phyletic gradualism. In: Schopf, T.J.M. (ed.) Models in Paleobiology, pp. 82–115. Freeman, Cooper, San Francisco, CA (1972)
Eliade, M.: The Sacred and the Profane: The Nature of Religion. Harcourt, Orlando, FL (1959)
Embree, J.: Review of the Chrysanthemum and the sword. Far East. Surv. **16**(1), 11 (1947a)
Embree, J.: Review of Hawaii's Japanese and the Chrysanthemum and the sword. Am. Sociol. Rev. **12**(2), 245–246 (1947b)
Embree, J.T.: Suye Mura: A Japanese Village. University of Chicago Press, Chicago, IL (1964 [1939])
Evans-Pritchard, E.: Witchcraft, Oracles, and Magic Among the Azande. Clarendon, Oxford (1976 [1937])
Ferguson, K.: Tycho and Kepler. Walker, New York (2002)
Feyerabend, P.K.: Against Method, 4th edn. Verso, New York (2010)
Feynman, R.P.: The Meaning of It All: Thoughts of a Citizen-Scientist. Perseus, New York (1998)
Firestein, S.: Ignorance: How It Drives Science. Oxford University Press, New York (2012)
Forsyth, A.: A Natural History of Sex: The Ecology and Evolution of Mating Behavior. Firefly, Richmond Hill, ON (2001 [1986])
Freud, S.: Totem and Taboo. Routledge and Kegan Paul, London (1950)
Geertz, C.: Religion as a cultural system. In: Geertz, C. (ed.) The Interpretation of Cultures, Selected Essays, pp. 87–125. Basic, New York (1973)
George, M.I.: Christianity and Extraterrestrials? A Catholic Perspective. iUniverse, New York (2005)
Gilkey, L.B.: Maker of Heaven and Earth: The Christian Doctrine of Creation in Light of Modern Knowledge. Doubleday, Garden City, NY (1965)
Glock, H.J.: Relativism, commensurability and translatability. In: Preston, J. (ed.) Wittgenstein and Reason, pp. 21–46. Blackwell, Oxford (2008)
Gorer, G.: Themes in Japanese culture. N. Y. Acad. Sci. Ser. II **5**, 106–124 (1943)
Gould, S.J., Eldredge, N.: Punctuated equilibria: the tempo and mode of evolution reconsidered. Paleobiology **3**(2), 115–151 (1977)
Grinspoon, D.: Lonely Planets. HarperCollins, New York (2009)
Hamilton, W.D.: The genetical evolution of social behaviour. J. Theor. Biol. **7**(1), 1–16 (1964)
Harris, M.: The Rise of Anthropological Theory. Harper and Row, New York (1968)
Hendershot, C.: Paranoia, the Bomb, and 1950s Science Fiction Films. Bowling Green State University Popular Press, Bowling Green, OH (1999)

Herzing, D.L.: SETI meets a social intelligence: dolphins as a model for real-time interaction and communication with a sentient species. Acta Astronaut. **67**(11–12), 1451–1454 (2010)
Hoyt, W.G.: Lowell and Mars. University of Arizona Press, Tucson, AZ (1996)
Hut, P.: Conclusion: life as a laboratory. In: Alan Wallace, B. (ed.) Buddhism and Science: Breaking New Ground, pp. 399–416. Columbia University Press, New York (2003)
Ikegami, E.: Shame and the samurai: institutions, trustworthiness, and autonomy in the elite honor culture. Soc. Res. **70**(4), 1351–1378 (2003)
Jermain, R., Kilgore, L., Trevathan, W.: Essentials of Physical Anthropology, 7th edn. Wadsworth, Belmont, CA (2009)
Jumain, R., Kilgore, L., Trevathan, W.R.: Essentials of Physical Anthropology. Wadsworth, Belmont, CA (2009)
Kapchan, D.A., Strong, P.T.: Theorizing the hybrid. J. Am. Folklore **112**(445), 239–253 (1999)
Kargel, J.S.: Mars: A Warmer, Wetter Planet. Springer, New York (2004)
Kennicutt Jr., R.C., Evans II, N.J.: Star formation in the Milky Way and nearby galaxies. Annu. Rev. Astron. Astrophys. **50**, 531–608 (2012)
Kincaid, H.: Defending laws in the social sciences. Philos. Soc. Sci. **20**(1), 56–83 (1990)
Kopf, G.: Beyond Personal Identity: Dōgen, Nishida, and a Phenomenology of No-Self. Curzon, Richmond, Surrey (2001)
Kuhn, T.: The Structure of Scientific Revolutions. University of Chicago Press, Chicago, IL (1962a)
Kuhn, T.: Historical structure of scientific discovery. Science **136**(3518), 760–764 (1962b)
La Barre, W.: Some observations on character structure in the orient: the Japanese. Psychiatry **8**, 319–342 (1945)
Lane, K.M.: Geographies of Mars. University of Chicago Press, Chicago, IL (2011)
Lemarchand, G.A., Lomberg, J.: Communication among interstellar intelligent species. In: Vakoch, D.A. (ed.) Communication with Extraterrestrial Intelligence, pp. 371–396. State University of New York Press, Albany, NY (2011)
Letaw Jr., H.: Cosmic storytelling: primitive observables as Rosetta analogies. In: Vakoch, D.A., Harrison, A.A. (eds.) Civilizations Beyond Earth: Extraterrestrial Life and Society, pp. 159–169. Berghahn, New York (2011)
Lowell, P.: Mars as the Abode of Life. MacMillan, New York (1908)
Margenau, H.: Reality in quantum mechanics. Philos. Sci. **16**(4), 287–302 (1949)
Margenau, H.: The Miracle of Existence. Ox Bow, Woodbridge, CT (1984)
Melville, R. (trans.): Lucretius: On the Nature of the Universe. Clarendon, London (1997)
Michaud, M.A.G.: Contact with Alien Civilizations. Copernicus, New York (2007)
Milbrath, S.: Star Gods of the Maya: Astronomy in Art, Folklore, and Calendars. University of Texas Press, Austin, TX (2010)
Minsky, M.: Communication with alien intelligence. In: Regis, E. (ed.) Extraterrestrials: Science and Alien Intelligence. Cambridge University Press, New York (1985)
Morgan, L.H.: Ancient Society. MacMillan, London (1877)
Morris, J.: Review of the Chrysanthemum and the sword. Pac. Aff. **20**(2), 208–210 (1947)
Morrison, P., Cocconi, G.: Searching for interstellar communications. Nature **184**(4690), 844–846 (1959)
Nagel, T.N.: What is it like to be a bat? Philos. Rev. **83**(4), 435–450 (1974)
Neal, M.: Preparing for extraterrestrial contact. Risk Manage. **16**(2), 6387 (2014)
Neufeld, M.: Von Braun: Dreamer of Space, Engineer of War. Alfred. A Knopf, New York (2007)
Oakley, B.: Concepts and implications of altruism bias and pathological altruism. Proc. Natl. Acad. Sci. U. S. A. (2013). doi:10.1073/pnas.1302547110
Oakley, B., Knafo, A., Madhavan, G., Wilson, D.S. (eds.): Pathological Altruism. Oxford University Press, New York (2012)
Ortner, S.: Anthropology and Social Theory: Culture, Power, and the Acting Subject. Duke University Press, Durham, NC (2006)
Paine, Thomas.: The Age of Reason. G.P. Putnam's Sons (1890)

Paley, W.: Natural Theology or Evidences of the Existence and Attributes of the Deity. J. Faulde, London (1802)
Pannenberg, W.: Toward a Theology of Nature: Essays on Science and Faith. Westminster/John Knox, Louisville, KY (1993)
Papagianni, D., Morse, M.A.: The Neanderthals Rediscovered: How Modern Science Is Rewriting Their Story. Thames and Hudson, New York (2013)
Pasquinelli, C.: The concept of culture between modernity and postmodernity. In: Hubinger, V. (ed.) Grasping the Changing World, pp. 53–73. Routledge, New York (1996)
Peterson, A.L.: Being Animal: Beasts and Boundaries in Nature Ethics. Columbia University Press, New York (2013)
Petigura, E.A., Howard, A.W., Marcy, G.W.: Prevalence of Earth-size planets orbiting Sun-like stars. Proc. Natl. Acad. Sci. U. S. A. **110**(48), 19273–19278 (2013)
Piszkiewicz, D.: Wernher von Braun: The Man Who Sold the Moon. Praeger, Westport, CT (1998)
Polkinghorne, J.: One World: The Interaction of Science and Theology. Templeton Foundation Press, Philadelphia, PA (2007 [1986])
Prashad, V.: Bruce Lee and the anti-imperialism of kung fu: a polycultural adventure. Positions **11**(1), 51–90 (2006)
Pring, H.: A transcendental account of correspondence and complementarity. In: Bitbol, M., Kerszberg, P., Petitot, J. (eds.) Constituting Objectivity: Transcendental Perspectives of Modern Physics, pp. 317–328. Springer, New York (2009)
Raglan: Review of the Chrysanthemum and the sword. Man **48**, 35 (1948)
Ryang, S.: Japan and National Anthropology: A Critique. RoutledgeCurzon, London (2004)
Sagan, C.: Contact. Pocket, New York (1985)
Schrödinger, E.: My View of the World. Ox Bow, Woodbridge, CT (1983)
Shostak, S.: What to do when ET calls? The Guardian. http://www.theguardian.com/education/2004/sep/09/research.highereducation6 (2004). Accessed 14 July 2014
Shostak, S.: Contact: what happens if a signal is found? Space.com. http://www.space.com/searchforlife/seti_whatif_060817.html (2006). Posted 17 August
Shuch, H.P. (ed.): Searching for Extraterrestrial Intelligence: SETI Past, Present, and Future. Praxis, Chichester, UK (2011)
Spencer, H.: Social Statistics: The Conditions Essential to Human Happiness Specified, and the First of Them Developed. D. Appleton, New York (1873)
Spencer, H.: Social Statics: The Conditions Essential to Human Happiness Specified, and the First of Them Developed. A. M. Kelly, New York (1969 [1851])
Tarter, J.C.: SETI and the religions of the universe. In: Dick, S. (ed.) Many Worlds: The New Universe, Extraterrestrial Life and the Theological Implications, pp. 143–149. Templeton Foundation Press, Philadelphia, PA (2000)
Tillich, P.: Systematic Theology, vol. 1. University of Chicago Press, Chicago, IL (1951)
Tomkins, S.: The Origins of Humankind. Cambridge University Press, Cambridge (1998)
Traphagan, J.W.: The Practice of Concern: Ritual, Well-Being, and Aging in Rural Japan. Carolina Academic Press, Durham, NC (2004)
Traphagan, J.W.: Altruism, pathology, and culture. In: Oakley, B., Knafo, A., Madhavan, G., Wilson, D.S. (eds.) Pathological Altruism, pp. 272–288. Oxford University Press, New York (2012)
Traphagan, J.W.: Rethinking Autonomy: A Critique of Principlism in Biomedical Ethics. State University of New York Press, Albany, NY (2013)
Traphagan, J.W., Brown, L.K.: Fast food and intergenerational commensality in Japan: New styles and old patterns. Ethnology **41**(2), 119–134 (2002)
Traub, W.A.: Terrestrial, habitable-zone exoplanet frequency from Kepler. Astrophys. J. **745**, 20 (2012). doi:10.1088/0004-637X/745/1/20
Trautmann, T.R.: Lewis Henry Morgan and the Invention of Kinship. University of California Press, Berkeley, CA (1988)
Tylor, E.B.: Primitive Culture: Researches into the Development of Mythology, Philosophy, Religion, Language, Art, and Custom, vol. 1. G. P. Putnam's Sons, New York (1920)

Vakoch, D.: Celestial music? Space.com, 22 December. http://www.seti.org/seti/projects/imc/articles/celestial_music.php (2000)
Vakoch, D.: Will ET's mathematics be the same as ours? Space.com, 11 January. http://www.seti.org/seti/projects/imc/articles/et_math.php (2001)
Vakoch, D.: Culture in the cosmos. Space.com. http://www.space.com/searchforlife/seti_culture_070503.html (2007). Posted 3 May
Vakoch, D.: An iconic approach to communicating musical concepts in interstellar messages. Acta Astronaut. **67**(11–12), 1406–1409 (2010)
Vakoch, D.A. (ed.): Archaeology, Anthropology, and Interstellar Communication. NASA SP-2013-4413. National Aeronautics and Space Administration, Washington, DC (2014a)
Vakoch, D.A. (ed.): Extraterrestrial Altruism: Evolution and Ethics in the Cosmos. Springer, New York (2014b)
Von Mises, R.: The definition of probability. In: Eagle, A. (ed.) Philosophy of Probability: Contemporary Readings, pp. 355–387. Routledge, London (2011)
Wagner, R.: The Invention of Culture: Revised Edition. University of Chicago Press, Chicago, IL (1981)
Ward, B.: Dr. Space: The Life of Wernher von Braun. Naval Institute Press, Annapolis, MD (2005)
Ward, P.D., Brownlee, D.: Rare Earth: Why Complex Life Is Uncommon in the Universe. Copernicus, New York (2003)
Watson, J.L. (ed.): Golden Arches East: McDonald's in East Asia. Stanford University Press, Stanford, CA (1997)
Wenseleers, T., Gardner, A., Foster, K.R.: Social evolution theory: a review of methods and approaches. In: Székely, T., Moore, A.J., Komdeur, J. (eds.) Social Behavior: Genes, Ecology, and Evolution. Cambridge University Press, New York (2010)
Wexler, B.E.: Brain and Culture: Neurobiology, Ideology, and Social Change. MIT Press, Cambridge (2006)
Whitehead, A.N.: Symbolism: Its Meaning and Effect. Macmillan, New York (1927)
Zafirovski, M.: The Enlightenment and Its Effects on Modern Society. Springer, New York (2011)

Index

J. Traphagan, *Extraterrestrial Intelligence and Human Imagination*, Space and Society, DOI 10.1007/978-3-319-10551-2

MIX
Papier aus verantwortungsvollen Quellen
Paper from responsible sources
FSC® C105338

If you have any concerns about our products,
you can contact us on
ProductSafety@springernature.com

In case Publisher is established outside the EU,
the EU authorized representative is:
Springer Nature Customer Service Center GmbH
Europaplatz 3, 69115 Heidelberg, Germany

Printed by Libri Plureos GmbH
in Hamburg, Germany